ALGORITHMS

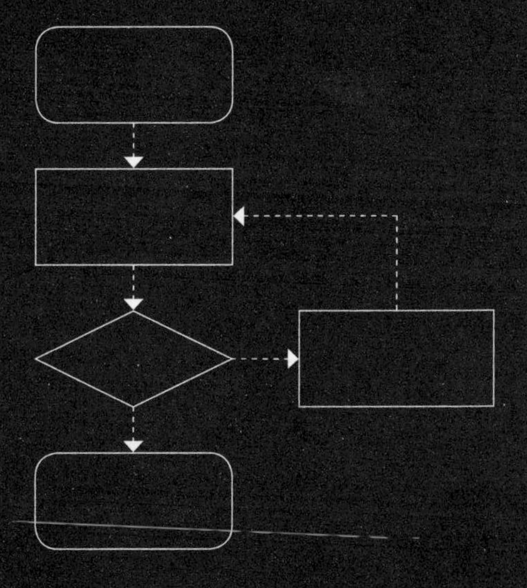

The MIT Press Essential Knowledge Series

A complete list of the titles in this series appears at the back of this book.

ALGORITHMS

PANOS LOURIDAS

The MIT Press | Cambridge, Massachusetts | London, England

This book was set in Chaparral Pro by New Best-set Typesetters Ltd. Printed and bound in the United States of America.

Library of Congress Cataloging-in-Publication Data

Names: Louridas, Panos, author.
Title: Algorithms / Panos Louridas.
Description: Cambridge, Massachusetts : The MIT Press, [2019] | Series: The MIT Press essential knowledge series | Includes bibliographical references and index.
Identifiers: LCCN 2019040771 | ISBN 9780262539029 (paperback)
Subjects: LCSH: Algorithms—Popular works. | Computer algorithms—Popular works.
Classification: LCC QA76.9.A43 .L668 2019 | DDC 005.13—dc23
LC record available at https://lccn.loc.gov/2019040771

10 9 8 7 6 5 4 3

The world is untranslatable but it is not incomprehensible, as long as you know the simple rule that nothing of what it expresses through its myriad lives and creatures is followed by a question mark, only by exclamation marks.

—Karl Ove Knausgaard, *Summer*

CONTENTS

SERIES FOREWORD

The MIT Press Essential Knowledge series offers accessible, concise, beautifully produced pocket-size books on topics of current interest. Written by leading thinkers, the books in this series deliver expert overviews of subjects that range from the cultural and the historical to the scientific and the technical.

In today's era of instant information gratification, we have ready access to opinions, rationalizations, and superficial descriptions. Much harder to come by is the foundational knowledge that informs a principled understanding of the world. Essential Knowledge books fill that need. Synthesizing specialized subject matter for nonspecialists and engaging critical topics through fundamentals, each of these compact volumes offers readers a point of access to complex ideas.

PREFACE

I know two young teenagers who possess more knowledge than any scientist, philosopher, or scholar of ages past. They are my sons. No, I am not a doting father who marvels at how extraordinarily gifted his children are. But these two kids have in their pockets devices that connect them with the vastest repository of information that has ever been created. There is no factual question they cannot answer, now that they have mastered the art of knowing where to look on the internet. They can translate from and to foreign languages without having to browse through hefty dictionaries—which we still keep in the house so that they know how things were, only a few years back. News, from anywhere, reach them in an instant. They can communicate with their peers before you know it, no matter where in the world they may live. They can plan their goings out in perfect detail. Alas, they can waste their time with abandon playing games or following trends that change so fast that I do not know why they matter.

All the above have become possible thanks to the huge advances in digital technology. Today we carry more computing power in our pockets than was used to ferry humans to the moon. As these two teenagers show, the changes in our lives have been immense; predictions for the future vary from utopias, where people will really not need to work, to dystopias, where the privileged few will lead fulfilling lives, with the rest being condemned to inconsequential torpor. Thankfully, we are able to shape this future, and an important factor in our ability to do this is how conversant we are with the technologies that underlie the achievements and the changes before us. Although we may lose sight of it in the bustle of our everyday lives, we live in the best period of human history. We are healthier than we have ever been, and expect to live longer, on average, than any generation that has ever lived. Despite the iniquity of glaring inequality, huge swathes of humanity have gotten rid of the shackles of poverty. We have never been closer to one another, both virtually and literally. We may decry the commercialism of mass global tourism, but cheap travel allows us to experience different cultures and visit places that we could once marvel about only in coffee table books. All this progress can and should continue.

To partake in this progress, however, it is not enough to use digital technology. We must be able to understand it. First, for the eminently practical reason that it offers excellent career opportunities. Second, because even if we don't care for a career in technology, we must know its underlying principles to appreciate its potential and shape our own role in it. Digital technology is enabled as much by its hardware, the physical components that make up computers and digital devices, as by its software, the programs that run on it. The backbone of programs are the algorithms that they implement: the set of instructions that describe the way to solve particular problems (if this does not look like a definition of what an algorithm is, don't worry, we have the rest of the book to fill out the details). Without algorithms, computers would be useless, and none of modern technology would exist.

What we need to know changes through time. For most of human history, schooling was not deemed necessary at all. Most people were illiterate, and if they were taught something, it would be mastery of some practical skill or scripture. In the beginning of the nineteenth century, more than 80 percent of the world's population was completely unschooled; now the vast majority has attained several years of school, and it is projected that by the end of the century, the proportion of unschooled people in the world will fall to zero. The years we spend on education have also increased. While in 1940 less than 5 percent of Americans had a bachelor's degree, by 2015 almost a third of them did.[1]

Back in the nineteenth century, no school would teach molecular biology because nobody knew anything about it; DNA wasn't discovered until well into the twentieth century. It now forms part of what we accept as the canon of an educated person's learning. Similarly, even though algorithms were discovered in antiquity, few people

Digital technology is enabled as much by its hardware, the physical components that make up computers and digital devices, as by its software, the programs that run on it. The backbone of programs are the algorithms that they implement.

troubled with them until the advent of modern computers. The author firmly believes that we have reached a point where algorithms are inside the core of what we consider to be essential knowledge. Unless we know what they are and how they work, we cannot understand what they can do, how they can affect us, what to expect from them, what their limits are, and what they require in order to work. In a society that increasingly functions thanks to algorithms, it behooves us as informed citizens to be knowledgeable about them.

It is also possible that learning algorithms helps us in another way. If learning mathematics introduces us to a way of rigorous reasoning, a familiarity with algorithms introduces us to a new way of algorithmic thinking: a way of reasoning to solve problems in a practical way so that efficient implementations of algorithms as programs can run fast in computers. The focus on designing processes that are practical and efficient can be a useful mental tool, even if we are not professional programmers.

This book aims to introduce algorithms to a nonspecialist audience in a way that the reader will understand how they really work. Its purpose is not to describe the effects of algorithms in our lives; there are other books that do a great job of depicting how improved processing of big data, artificial intelligence, and the weaving of computing devices into the fabric of our everyday lives may change the human condition. Here we are not interested

A familiarity with algorithms introduces us to a new way of algorithmic thinking: a way of reasoning to solve problems in a practical way so that efficient implementations of algorithms as programs can run fast in computers.

in *what* may happen but rather the *how* this can happen. To do that, we'll present real algorithms and show not only what they do but also how they actually function. Instead of hand waving, we'll provide detailed explanations.

To the question, "What are algorithms?" the answer is surprisingly simple. They are particular ways to solve our problems. These ways to solve our problems can be described in easy steps so that computers can execute them with amazing speed and efficiency. Yet there is nothing magical about these solutions. The fact that they comprise simple elementary steps means that there is no reason why they should be beyond the grasp of most people.

Indeed, the book does not assume knowledge of material beyond that commonly taught in high schools. Some mathematics does appear in the following pages because you cannot talk seriously about algorithms without *some* notation. Any concepts that are commonplace in algorithms but are not that common outside computer science are explained in the text.

The late physicist Stephen Hawking wrote in the introduction of his best-selling book *A Brief History of Time*, published in 1988, "Someone told me that each equation I included in the book would halve the sales." This sounds pretty ominous for the present book because mathematics does occur more than once. Yet I decided to press ahead, for two reasons. First, while the level of mathematics required for Hawking's physics is at the university level or beyond, the mathematics presented here is much more accessible. Second, as the purpose of this book is to show not just what algorithms are for but how they really work too, the reader should get to share some of the vocabulary we use when we discuss algorithms. And this vocabulary does include some mathematics. The notation is not the prerogative of the technical clerisy, and familiarity with it will help dispel any mystique surrounding the subject; in the end, we'll see that it mostly comes down to being able to talk about things in a precise quantitative way.

It is impossible to cover the whole subject of algorithms with a book like this, but it is possible to provide an overview and introduce a reader to algorithmic thinking. The first chapter lays the ground by introducing what algorithms are and how we can gauge their efficiency. We can say at the outset that an algorithm is a finite sequence of steps that we can perform with a pen and paper, and this plain definition would not be far from the truth. Chapter 1 starts from there, while also exploring the relationship between algorithms and mathematics. A key difference between the two is practicality; in algorithms, we are interested in practical ways to solve our problems. This means that we need to be able to measure how practical and efficient our algorithms are. We'll see that these questions can be carefully framed through the notion of computational complexity; this will inform the discussion of algorithms in the rest of the book.

The next three chapters look at three of the most essential application areas of algorithms. Chapter 2 covers algorithms that deal with the solution of problems relating to networks, called graphs, of things. These problems may include finding your way in a road network or sequence of links connecting you to somebody on a social network. They also include problems in other areas that are not immediately obvious in terms of their relationship: DNA sequencing and scheduling tournaments; this will illustrate that distinct problems can be solved efficiently using the same tools.

Chapter 3 and chapter 4 explore how to search for things and put things in order. These may seem prosaic, yet they are among the most important applications of computers. Computers spend a lot of time sorting and searching, but we are largely oblivious to this fact exactly because they are an integral, invisible part of most applications. Sorting and searching also offer us a glimpse of an important facet of algorithms. For many problems, we know of more than one algorithm to solve them. We choose among the available algorithms based on their particular characteristics; some algorithms are more suitable for certain problem instances than others. It is therefore instructive to see how different algorithms, with different characteristics, go about solving the same problem.

The following two chapters present important applications of algorithms on a large scale. Chapter 5 picks up graphs again to explain the PageRank algorithm, which can be used to rank web pages in order of significance. PageRank was the algorithm used by Google when it was founded. The success of the algorithm at ranking web pages in search results played a critical role in the phenomenal success of Google as a company. Fortunately, it is not difficult to grasp how PageRank works. It will give us the opportunity to see how an algorithm can solve a problem that on first impression, does not seem amenable to a computer solution: How do we judge what is important?

Chapter 6 introduces one of the most active areas in computer science: neural networks and deep learning. Successful applications of neural networks are reported in popular media. Stories pique our interest by describing systems that perform tasks such as image analysis, automatic translation, or medical diagnosis. We'll start out simple, from individual neurons, building up bigger and bigger neural networks that are able to perform more and more complex tasks. We'll see that they all work based on some fundamental principles. Their efficacy rises from the interconnection of many simple components and the application of an algorithm that lets neural networks learn.

After sketching what algorithms can do, the epilogue explores the limits of computation. We know that computers have performed amazing feats and expect so much more from them in the future, yet are there things that they cannot do? The discussion of the limits of computing will allow me to offer a more precise explanation of the nature of algorithms and computing. We said that we could describe it as a finite sequence of steps that can be performed with a pen and paper, but what kind of steps could these be? And how close is the pen-and-paper analogy with what algorithms really are?

ACKNOWLEDGMENTS

First and foremost, I am grateful to Marie Lufkin Lee at the MIT Press for coming up with the idea for this book, Stephanie Cohen for goading me gently through the process, Cindy Milstein for her meticulous editing, and Virginia Crossman for her excellent attention to detail and taking care of everything. A book on algorithms should be part of the Essential Knowledge series, and I am proud that I am the one to write it.

I extend my thanks to Diomidis Spinellis for commenting on parts of the book, and my special appreciation to Konstantinos Marinakos, who read the manuscript, spotted embarrassing bugs, and offered generous suggestions for improvements.

Finally, I want to express my gratitude to two teenagers, Adrianos and Ektor, whose lives will to such an extent be determined by the subject matter of this book, and their mother, Eleni; they enabled me to make this happen.

1

WHAT IS AN ALGORITHM?

The Algorithmic Age

We like putting labels on time periods, perhaps because affixing a tab on time allows us to get a grip on its fluidity. We have therefore started speaking of the present as the dawning of a new *algorithmic age*, in which algorithms will reign supreme, and will govern larger and larger parts of our lives. It is interesting that we are not talking about the *computer age* or *internet age* anymore. We somehow take them for granted. It is when we add algorithms that we begin intimating that perhaps something qualitatively different has started taking place. "Behold the Almighty Algorithm, a snippet of computer code coming to stand for a Higher Authority in our secular age, a sort of god," says Christopher Lydon, former *New York Times* journalist and host of the *Radio Open Source* show. And indeed, algorithms are taken to be some form of higher authority when they are used to organize political campaigns, follow our traces in the online realm, shadow our shopping and target us with advertising, suggest dating partners, or monitor our health.[1]

There is an aura of mystery around all that, which perhaps flatters the acolytes of algorithms. Being described a "programmer" or "computer scientist" marks you as a decent, albeit somewhat technical, character. How much better to be a member of the tribe that is about to change almost everything in our lives?

There is definitely a sense in which algorithms are a sort of god. They are mostly held unaccountable, like gods; things happen, not because of human agency, but because they were decided by an algorithm, and the algorithm sits beyond the pale of responsibility. Machines, running algorithms, can surpass human performance in more and more fields so that it appears that the area of human superiority is reduced day by day; some believe that the day where computers will be able to surpass humans in every aspect of cognition is not far away.

But there is also a sense in which algorithms are nothing like gods, although we often lose sight of it. An algorithm does not produce its results by an act of revelation. We know exactly the rules that it follows and kinds of steps it takes. No matter how wonderful the outcome, it can always be traced back to some elementary operations. To people who are newcomers to algorithms, it may come as a surprise how elementary these may be. That is not to besmirch algorithms; seeing how something really works may take out some part of its mystique. At the same time, understanding how something works may allow us to appreciate the elegance of its design, even if it is no longer mysterious.

The premise of this book is that indeed algorithms are not mysterious. They are tools that allow us to do certain things well; they are specific kinds of tools whose purpose is to allow us to solve problems. In this way they are cognitive tools; as such, they are not the only ones. Numbers and arithmetic are also cognitive tools. It took us thousands of years to evolve a number system that children can learn in school so that they can perform calculations that would be impossible without it. Now we take numeracy for granted, but a few generations back only a small minority of humans had any knowledge of it.

Similarly, knowledge of algorithms should not be the prerogative of a small elite minority; as cognitive tools they can be apprehended by all kinds of people, not just computer professionals. What is more, they *should* be understood by more people because that will allow us to put algorithms into perspective: to know what they do, how they do it, and what we can realistically expect them to do.

An essential knowledge of algorithms is what we are after here so that we can take a meaningful part in the conversations on the algorithmic age. That is not an age that is thrust on us, but one of our own creation, based on tools that we have devised. The study of these tools is the subject of this book. Algorithms are beautiful tools, and a glimpse of how they are made and work can enhance our way of thinking.

We'll start by dispelling an irksome notion: that algorithms are about computers. This, we'll see, makes as much sense as saying that numbers are about calculators.

A Way to Do Things

A pen-and-paper puzzle, music, divisibility of numbers, and neutron accelerators in particle physics—we'll see that what they all have in common is the same algorithm, applied to such different domains, yet working on the same underlying principles. How can this be?

The word "algorithm" itself does not reveal its meaning. It comes from the name of Muḥammad ibn Mūsā al-Khwārizmī (ca. 780–ca. 850), a Persian scholar who worked on mathematics, astronomy, and geography. Al-Khwārizmī's contributions were many and widespread. The term "algebra" comes from the Arabic title of his most influential work, *The Compendious Book on Calculation by*

Completion and Balancing. His second most influential book, *On the Calculation with Hindu Numerals*, was on arithmetic and, translated into Latin, introduced the Hindu-Arabic numeral system to the West. Al-Khwārizmī's name was latinized to *Algorismus*, which came to denote the method of numerical computation with the decimal numbers. Algorismus, influenced by the Greek word for "number" (*arithmos*, as in arithmetic), became algorithm, still denoting decimal arithmetic, before acquiring its modern sense in the nineteenth century.

You could be tempted to think that algorithms are something that we do with computers, but this would be wrong. It is wrong because we had algorithms long before we had computers. The first-known algorithms date back to ancient Babylon.[2] It is also wrong because algorithms are not about problems that have to do with computers. Algorithms are about doing something in a specific way, following some kind of steps. That is somewhat vague. You may ask, What kind of steps? What specific way? We can dismiss all vagueness, and give a precise mathematical definition of what an algorithm is and what it does—such a definition does exist—but we don't need to go to such lengths. You may be happy to know that an algorithm is a set of steps that you can follow with pen and paper, and you can be assured that this seemingly facile description is close to those used by mathematicians and computer scientists.

You could be tempted to think that algorithms are something that we do with computers, but this would be wrong. It is wrong because we had algorithms long before we had computers.

So we can start our approach to algorithms with a problem that we can solve by just writing things down. Suppose we have two sets of objects and want to spread the objects of one of the two sets as evenly as possible among the objects of the other set. We will use crosses (×) for the objects of the first set and bullets (•) for the objects of the second set. We want to spread out crosses among the bullets.

If the number of crosses divides the total number of objects, that is easy. We just partition the crosses among the bullets as if we would do division. For example, if we have 12 objects in total, out of which three are crosses and nine are bullets, we put one cross, then three bullets, then one cross, three bullets, and finally another cross and three bullets:

× • • • × • • • × • • •

But what if the total number of objects, crosses and bullets taken together, cannot be divided exactly by the crosses? What can we do if we have five crosses and seven bullets?

We start by putting all crosses followed by all bullets in one row as follows:

× × × × × • • • • • • •

Then we take five bullets and place them under the crosses:

× × × × × • •

• • • • •

We notice in the pattern that emerges that we have a remainder of two columns to the right. We take the two remainder columns, each comprising a single bullet, and put them under the first two columns, forming a third row:

× × × × ×

• • • • •

• •

Now we notice that we have a remainder of three columns. We take the rightmost two of them and put them under the two leftmost columns:

× × ×

• • •

• •

× ×

• •

Now we have only one remainder column, so we stop. We concatenate the columns from left to right and get:

× • • × • × • • × • × •

This is the result. We have distributed the crosses among the bullets. They are not as evenly spaced as before, but that is impossible to do because, remember, five does not divide evenly into 12. We have managed to avoid heaping all the crosses together, however, and have created a pattern that does not look entirely haphazard.

You may wonder if there is anything particular about this pattern; it helps if you substitute DUM for the cross and da for the bullet. Then the pattern goes DUM-da-da-DUM-da-DUM-da-da-DUM-da-DUM-da and it really is a rhythm. A rhythm is constituted by accented parts, also called *onsets*, and unaccented or silent parts. The rhythm we found is not a rhythm of our own devising. It is used by the Aka pygmies in the Central African Republic; it is the clapping, called Venda, of a South African song; it is also a rhythm pattern used in Macedonia, in the Balkans. There is more. If we rotate it, so that it starts at the second cross (that is, onset), then it becomes:

× • × • • × • × • × • •

That is the Columbia bell pattern, popular in Cuba and West Africa, as well as a drumming pattern in Kenya, while it is also used in Macedonia (again). If we rotate it to start on the third, fourth, and fifth onset, other popular rhythms around the world emerge.

Is this just a one-off thing? We can try to create a 12-part rhythm out of seven onsets and five silent parts—kind of mirroring the five onsets and seven silent parts that we had before. If we follow exactly the same procedure, we will arrive at:

× • × × • × • × × • × •

This, again, is a rhythm. It is used in the Mpre rhythm of the Ashanti in Ghana, and if we start it on the last onset, it is used by the Yoruba in Nigeria as well as in Central Africa and Sierra Leone.

Lest you think we have geographic omissions, if we start with five beats and 11 silent parts, we arrive at the following:

× • • × • • × • • × • • × • • •

That is the Bossa-Nova rhythm, rotated. The actual Bossa-Nova rhythm starts on the third onset, so the exact correspondence is:

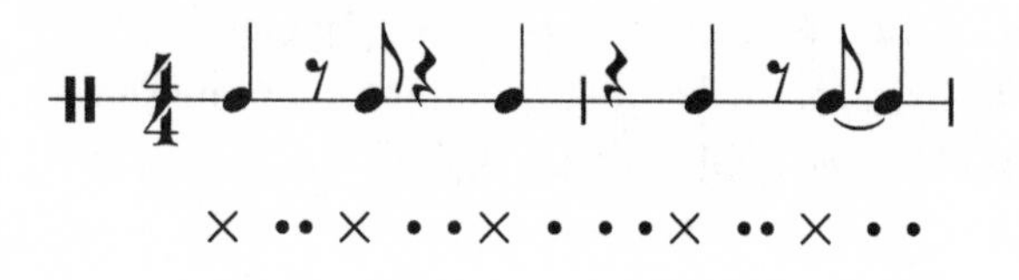

If we try with three beats and four silent parts, we get the pattern:

× • × • × • •

This rhythm in a seven/four meter is popular, and not just in traditional music. Among other tunes, it is the rhythmic pattern of Pink Floyd's song "Money":

Many more rhythms can be derived in this way by putting crosses and bullets in columns, and moving them around in the way we just described. We illustrated the procedure by measuring remainder columns, but this really is a pictorial way of showing what really happens. Instead of creating columns, checking the geometry, and moving them around, we can do the same thing more formally with simple numerical operations. To see what, let's return to the example of 12 parts and seven onsets. We start by dividing 12 by 7, which gives us quotient 1 and remainder 5:

$$12 = 1 \times 7 + 5$$

This tells us to put the seven onsets in the beginning, creating seven columns of onsets, followed by a remainder of the five unaccented parts:

× × × × × × × • • • • •

Now we divide again, but this time we divide the divisor of the previous division, 7, by the remainder of the previous division, 5. This gives us a quotient of 1 again while the remainder is 2:

$$7 = 1 \times 5 + 2$$

This means that we need to take the five rightmost columns and place them under the five leftmost columns, leaving a remainder of 2:

× × × × × × ×
• • • • •

We repeat the same step: we divide the divisor of the previous division, 5, by the remainder of the previous division, 2. The quotient is 2 and the remainder is 1:

$$5 = 2 \times 2 + 1$$

This tells us to take *twice* the two rightmost columns and place them under the two leftmost columns, leaving a remainder of 1:

× × ×
• • •
× ×
× ×
• •

Note that *twice* means that this is equivalent to what we would be doing in two steps if we had worked as we were doing before, without using the division. We would go from:

× × × × × × ×
• • • • •

first to:

× × × × ×
• • • • •
× ×

and then to:

× × ×

• • •

× ×

× ×

• •

If we concatenate the columns, we get the Mpre rhythm:

× • × × • × • × × • × •

Our First Algorithm

We can write down the method we followed in a bit more precise terms as the following steps. We assume that we start with two numbers, *a* and *b*. We let *a* be the total number of parts. If the number of onsets is greater than the number of the silent parts, then *b* is the number of onsets. Otherwise it is the number of the silent parts. At the beginning, we create a row with the onsets followed by the silent parts.

1. Perform the division of *a* by *b*. This will give us a quotient and remainder. If we call the quotient *q* and remainder *r*, we'll have $a = q \times b + r$. This is integer division as we know it. We take *q* times the rightmost

b columns and move them under the leftmost columns, leaving a remainder of *r* columns on the right.

2. If the remainder *r* is equal to zero or one, then we stop. Otherwise, we go back to step 1, but this time *b* will be the new *a* and *r* will be the new *b*. Or in other words, we go back to step 1, setting *a* equal to *b* and *b* equal to *r*.

In these two steps we perform a division repeatedly, until it does not make sense to repeat it. You can trace the steps we take in the following table, where we start with $a = 12$ and $b = 7$, like we did before; in each row we have $a = q \times b + r$:

a	q	b	r
12	1	7	5
7	1	5	2
5	2	2	1

If you examine the table, you can verify that each row corresponds to one step of the column formation and moving, but we have a more precise definition of the method we used. In fact, we have a series of steps that we can perform with pen and paper, so this is our first algorithm! We have an algorithm for creating patterns that correspond to many, and indeed surprisingly many, musical rhythms. Working with different numbers of offsets and silent parts, we can get about 40 rhythmic patterns that are found in different rhythms around the world. That should give us pause for a minute: it is a simple algorithm (only two steps, repeated) and yet able to produce so many interesting results.

Our algorithm does more than that, though. As we are talking about the division of two numbers, let us consider the following general problem: If we have two numbers *a* and *b*, what is the greatest number that divides them both? This is called the *greatest common divisor*, or *gcd*, of the two numbers. We encounter the greatest common divisor in elementary arithmetic, in problems such as, If we have 12 packets of crackers and four packets of cheese, how will you distribute them in baskets so that you have the same proportion of crackers and cheese in each basket? As four divides 12, you will have four baskets, each containing three packets of crackers and one packet of cheese; the greatest common divisor of 12 and four is four. Things get more interesting if you have 12 packets of crackers and eight packets of cheese. You cannot divide one by the other, but the greatest number that divides both 12 and eight is four, which means that you will make again four baskets, each containing three packets of crackers and two packets of cheese.

So how can we find the greatest common divisor of any two integer numbers? We have seen that if one of the numbers divides the other, that is the greatest common divisor.

But if that does not happen, then it turns out that in order to find the greatest common divisor of two numbers, we only need to find the greatest common divisor of the remainder of the division of the two numbers and the second number. This is actually easier to see with symbols. If we have two integers a and b, the greatest common divisor of a and b is equal to the greatest common divisor of the remainder of $a \div b$ and b. This brings us back to our rhythms. *The way we have been finding rhythms is in fact the same way we use to find the greatest common divisor between two numbers.*

The way to find the greatest common divisor between two numbers is called *Euclid's algorithm*, in honor of Euclid, an ancient Greek mathematician who first described it in his books *Elements* (ca. 300 BCE). The basic idea is that the greatest common divisor between two numbers remains the same if we replace the larger number of the two with its difference with the smaller number. Take 56 and 24. Their greatest common divisor is 8, which is also the greatest common divisor of $56 - 24 = 32$ and 24, and the same goes for 32 and 24, and so on. Repeated subtraction is really division, so Euclid's algorithm is described with the following steps:

1. To find the greatest common divisor of a and b, perform the division of a by b. This will give us a quotient and remainder. If we call the quotient q and remainder r, we'll have $a = q \times b + r$.

2. If the remainder r is equal to 0, then we stop, and the greatest common divisor of a and b is b. Otherwise, we go back to step 1, but this time b will be the new a and r will be the new b. Or in other words, we go back to step 1, setting a equal to b and b equal to r.

These are essentially the same steps as before. The only difference is that when finding rhythms, in step 2 we stop when the remainder is 0 or 1, while Euclid's algorithm stops when the remainder is 0. This is really the same: if you have a remainder of 1, then in the next repetition of step 1, you get a 0 remainder because 1 divides every integer. Try 9 and 5: $9 = 1 \times 5 + 4$, so we go to $5 = 1 \times 4 + 1$ and then $4 = 1 \times 4 + 0$, so the greatest common divisor of 9 and 5 is 1.

It may help you to see the algorithm in action with $a = 136$ and $b = 56$ in the following table, similar to the one we saw before with our rhythms. We find that the greatest common divisor of 136 and 56 is the number 8:

a	q	b	r
136	2	56	24
56	2	24	8
24	3	8	0

As we noted with 9 and 5, Euclid's algorithm works correctly in all cases, even when the two numbers do not have any common divisor apart from 1. This is what happened with $a = 9$ and $b = 5$. You can see for yourself what happens if you try to perform the algorithm's steps with $a = 55$ and $b = 34$; it will take a few steps, but the algorithm will determine that the only common divisor is 1.

The steps in Euclid's algorithm are performed in a well-defined order. The description of the algorithm illustrates the way its component steps are combined:

1. The steps are put in a *sequence*.

2. Steps may describe a *selection* that determines which steps to follow. In step 2, there is a test of whether the remainder is 0 or not. Then there are two alternatives, depending on the outcome: we either stop or go back to step 1.

3. Steps can be put into a *loop* or *iteration*, where they are executed repeatedly. In step 2, if the remainder is not equal to 0, we go back to step 1.

We call these three ways to combine steps *control structures* because they dictate which action will be performed as we carry out the algorithm. All algorithms are structured in this way. They comprise steps doing calculations and processing data; these steps are assembled together and choreographed using these three control structures. More complex algorithms have more steps, and their choreography may be more complex. But the three control structures suffice to describe the way the steps of any algorithm should be put together.

The steps of an algorithm will, among other things, operate on the input we provide. The input is the data that are processed by the algorithm. If we adopt a data-centric view, we use an algorithm to transform some data, which describe a problem, to some form that corresponds to the problem's solution.

We found an algorithm behind musical rhythms that is an application of division, but in reality, we need not look that far; the act of division itself is an algorithm. Even if you have not heard of Euclid's algorithm, you know how to divide two large numbers; we have all spent time in our early years learning to perform long multiplication and long division. Our teachers spent hours drilling into our heads how to perform these operations: a set of steps for putting numbers in the right places and doing things with them—they are algorithms. But algorithms are not simply about numbers, as we have just seen. We just found that they are about how we can produce music. Yet there is nothing mystifying about that. A rhythm is a way to distribute stresses in a time interval, and the same principle is at work when we pack crackers and cheese.

The application of Euclid's algorithm to rhythms had an unlikely source: a *neutron source* facility in the Oak Ridge National Laboratory in Tennessee. The Spallation

Neutron Source (SNS) there produces intense pulsed neutron beams that are used in experiments in particle physics. (The verb *to spall* means breaking a material into smaller pieces; in nuclear physics, we have a heavy nucleus emitting a large number of protons and neutrons after being bombarded with a high-energy particle.) In the operation of the SNS, some components, such as high-voltage power supplies, should run so that pulses are distributed in timing slots as evenly as possible. An algorithm devised to do the distribution is essentially the same as the rhythm-making algorithm and Euclid's algorithm, taking us from numbers to subatomic particles to music.[3]

Algorithms, Computers, and Mathematics

We said that algorithms are not about computers, yet today most people bundle them together. It is true that algorithms show their potential when they are coupled with computers, but a computer is really a machine with the special trait that we can order it to do certain things. We order it by *programming* it, and usually we program it to execute algorithms.

Which brings us to programming itself. Programming is the discipline of translating our intentions to some notation that a computer is able to understand. We call this notation a *programming language* because sometimes

Programming is the discipline of translating our intentions to some notation that a computer is able to understand. We call this notation a *programming language*.

it does look like we are writing in a human language, but programming languages are fairly simple affairs compared to the richness and complexity of human languages. Now, of course, a computer does not really understand anything. Things may change in the future, if we are able to produce truly intelligent machines, but right now when we say that a computer understands a notation, it really means that the notation is converted to a series of instructions for manipulating current in electronic circuits (we may also use light instead of electric current, yet the idea is the same).

If an algorithm is a set of steps we can carry out ourselves, programming is the activity by which we write down the steps in the notation that the computer understands. Then it is the computer that will carry them out. Computers are much faster than human beings, so they can execute the steps in less time. The fundamental factor in computing is *speed*. A computer cannot do something qualitatively different from what we humans can do, but it can do it faster—a lot faster. An algorithm gains power on a computer because it can be executed there in a fraction of the time it would take us to perform the same steps, *but they are still the same steps*.

A programming language gives us a way to describe to a computer the steps of algorithms. It also provides the means to structure them using the three fundamental control structures: sequence, selection, and iteration. We

If an algorithm is a set of steps we can carry out ourselves, programming is the activity by which we write down the steps in the notation that the computer understands.

write the steps and describe how they are choreographed using the vocabulary and syntax provided by the particular programming language we are using.

There is an additional advantage to using computers apart from speed; if you can recall how you learned to perform long multiplication and division, it may have taken a lot of practice, and may not have been that exciting. As we noted above, these things are drilled into our heads at an early age, and drilling inside a head is not a pleasant procedure. Computers do not suffer from boredom, so an added reason to have them perform algorithms is to take out the tedium and leave us time to do more interesting things.

Although an algorithm is usually executed on a computer, after being written in a programming language, it is primarily written for humans, who must understand how it works and when it can be used. This brings us to something essential that even experienced computer scientists and seasoned programmers forget. The only way to truly understand an algorithm is to perform it by hand. We must be able to execute the algorithm, in the same way the computer would execute a program that implements it. At this date and time, we are privileged to have at our disposal an amazing array of media that can help us learn: superb videos, animations, and simulations are one click away. All these are great, but when you are stuck, have your pen and pad nearby. The same applies to these very lines. Have you really understood how you can create rhythms?

Did you try to create one? Can you find the greatest common divisor of 252 and 24?

All programs implement a set of steps to do something, so we could be tempted to say that all programs are algorithms. We are a bit stricter, however, and want our steps to meet certain characteristics:[4]

1. The steps must terminate after a finite number of steps. An algorithm cannot run forever. (A program may run forever, as long as the computer on which it runs remains operational. That program would not be an implementation of an algorithm; it would just be a computational process.)

2. The steps must be precise, so that we can execute them without confusion.

3. The algorithm may operate on some input; in the case of Euclid's algorithm, it operates on two integers.

4. The algorithm has some output; that is the whole purpose of the algorithm: to produce something as a result. In Euclid's algorithm, that is the greatest common divisor.

5. The algorithm must be effective. A human should be able to execute each step in a reasonable amount of time with pen and paper.

These characteristics ensure that the algorithm does something. An algorithm exists because it does something useful. Frivolous algorithms do exist, and computer scientists may invent useless algorithms either in jest or by mistake, but we are really interested in algorithms that have some utility to us. When working with algorithms, it is not enough to show that something can be done. We want algorithms to be of practical interest, and for that purpose they must do something well.

Therein lies a fundamental difference between algorithms and mathematics. Most early computer scientists were mathematicians, and computer science uses a lot of mathematics, but it is not a mathematical discipline. A mathematician wants to prove that *something is so*; a computer scientist wants to *make it work*.

Our first characteristic of an algorithm is that it should require a finite number of steps. That is not very precise. We do not want to have just a finite number of steps. We want to have a number of steps that is small enough to execute them in practice, so that our algorithm finishes in a reasonable amount of time. That means that coming up with an algorithm is not enough; the algorithm must also be effective in practice. Let's see an example to illustrate the difference between knowing something and knowing how to do something efficiently. Imagine we have a grid like the following:

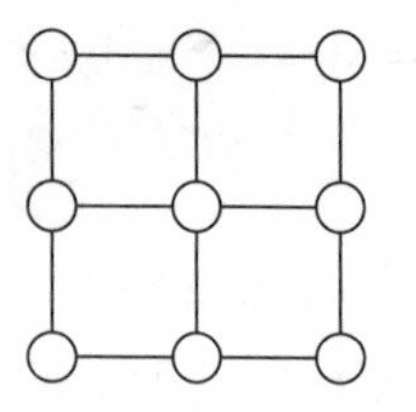

We want to find the shortest path from the upper-left corner of the grid to the lower-right corner, without visiting the same place twice. The length of each path is equal to the number of links between points on the grid. Here is one way to do it: find all such paths, measure how long each of them is, and pick up the shortest, or any of the shortest in case of ties. The total number of paths is 12, which you can see below:

There are five paths of length 4, so we can pick any one of them.

We are not limited to 3×3 grids, though. We can have 4×4, 5×5, and even larger grids. Then we discover that our method does not scale well. There are 184 paths from the upper-left corner to the bottom-right corner of a 4×4 grid; if we go to the 5×5 grid, the number of such paths increases to 8,512. The number of paths continues to increase apace—in fact, at ever larger paces—and even counting such paths is a challenge. When we reach a 26 × 26 grid, we get 8 402 974 857 881 133 471 007 083 745 436 809 127 296 054 293 775 383 549 824 742 623 937 028 497 898 215 256 929 178 577 083 970 960 121 625 602 506 027 316 549 718 402 106 494 049 978 375 604 247 408 paths. This number has 151 decimal digits and was found with a computer program implementing an algorithm; yes, we use an algorithm to understand the behavior of another algorithm.[5]

The procedure for enumerating all paths and picking the shortest one is undoubtedly correct, and will always give us the shortest path—or any of the shortest paths, if there are many equally short ones—yet it is definitely impractical. Also, it is completely useless, as there are algorithms that will find the shortest path without having to enumerate all possible paths, thus saving a lot of time and allowing us to tackle grids of any size. In the 26 × 26 grid, the number of steps required to find the answer is only in the order of the hundreds; we'll see it in the next chapter.

The question of what is a practical algorithm and in what sense an algorithm is more practical than others is at the heart of any application of them. We'll see in the rest of the book that there often exist different algorithms for solving the same problem and we choose the algorithm that is most appropriate for the application at each particular setting. Like all tools, some algorithms are more suitable for particular cases than others. Unlike many other tools, though, we possess a well-defined way to evaluate the merits of algorithms.

Measuring Algorithms

When we are investigating an algorithm to solve a problem, we want to know how it is going to perform. Speed is always an important factor. We use algorithms on computers to do things faster than a human would do.

As computer hardware improves, we are usually not content with knowing how a program implementing an algorithm runs on a particular computer. Our computer may be faster or slower than the one that the algorithm was measured on, and after some years, measurements of algorithms on outdated machines will have only historical interest. We need a way to measure how well an algorithm performs independent of computer hardware.

The size of the problem we are trying to solve, though, should be somehow reflected in how we measure the performance of an algorithm. We don't really care how long it takes to sort 10 items; after all, we can do that by hand. We care how long it takes to sort a million items or more. We want a measure of how we expect an algorithm to perform in problems that are not trivial.

To do that, we need a way to quantify the size of problems fed to algorithms. The dimension of interest varies among different problems. If we want to sort a number of items in our computer, the relevant dimension is the number of items that we want to sort (and not, say, the size or composition of the items). If we want to multiply two numbers, the relevant dimension is the number of digits of the two numbers (that also makes sense for humans: long multiplication is *long* because it depends on how many digits each number has). When we study a problem and candidate algorithms for tackling it, we do it always with the size of the problem under consideration.

Although particular problems have different ways to assess their size, in the end, for each problem we specify its size with an integer number, which we call n. Picking up the examples above, n is the number of either the items to sort or digits of the numbers we want to multiply. Then we want to be able to talk about the performance of algorithms tackling problems of size n.

The time required by an algorithm is related to its *computational complexity*. The computational complexity of an algorithm is the amount of resources it requires to run. There are two main kinds of resources here: time, how long it takes, and space, how much storage it requires in terms of computer memory.

We are focusing on time right now. As there are computers with different performance characteristics, talking about the time taken by an algorithm to run on a particular computer may give us some indication of what to expect when it runs on other computers, but we would like something more general. The speed of a computer depends on the time it takes to execute basic operations. To get around such specificities, we instead choose to talk about *the number of operations* required to run an algorithm, not the actual time it takes on a specific computer to run these operations.

Now having said that, note that we'll be abusing terminology a bit and treating "operations" and "time" as synonyms. Although we should be strictly saying that an algorithm requires "x operations," we'll also be saying that the algorithm is "time x," to indicate that it runs in the time required to execute x operations on any computer that the algorithm is actually run. Even though the actual time will vary with different hardware, it does not matter when we want to compare two algorithms that run on "time x" and "time y" on the *same* computer, whatever computer that is.

Now we return to the size of the problem given to an algorithm. As we are interested in nontrivial problems, we won't care about what happens with small problem sizes. We will be concerned with what happens once we reach a certain size. We won't say exactly what this size is, but we will always assume that it is substantial.

There is a definition of complexity that has proved to be useful in practice. It also has a symbol and name. We write $O(\cdot)$ and call it the *big O* notation. Inside the big O, in the place of the dot, we write an expression. The notation means that the algorithm will take time that is at most a multiple of the expression. Let us see what that means:

- If you want to look for something in a sequence of items—there are n items—and the sequence of items is not ordered in any way, the complexity is $O(n)$. That is, for n items, the time required to find a particular one in them will not be more than a multiple of the number of items.

- If you want to multiply two n digit numbers using long multiplication, the complexity is $O(n^2)$. That is, the time required for the multiplication will not be more than a multiple of the square of the size of the numbers.

If we have an algorithm that has $O(n)$ complexity, then for an input size of 10,000 we expect it to need a multiple of ten thousand steps. If the algorithm has $O(n^2)$ complexity, for a similarly sized input, we expect it to need a hundred million steps. For many problems, this is not a large size. Computers routinely sort 10,000 items. But you see that the scale of the number of steps represented by the algorithm's complexity can grow large.

Here are some examples that may help you appreciate the size of some numbers that we will encounter. Take the number 100 billion, or 10^{11}; this is one with 11 zeros behind it. If you take 100 billion hamburgers and lay them end to end, you can circle the earth 216 times, go to the moon, and come back.

A billion of something is usually called *giga* something, at least in computers. Next after the billion, or giga, comes the trillion, or *tera*, which is 1,000 billion, 10^{12}. If you start counting one number per second, you will need 31,000 years to get to one trillion. Up by 1,000 again and we get to one quadrillion, 10^{15}, or *peta*; the total number of ants that live on the earth is between 1 and 10 quadrillion, according to biologist E. O. Wilson. In other words, we have between 1 and 10 petaants on our planet.

After quadrillion comes quintillion, or *exa*; a quintillion is 10^{18} and is about the number of grains of sand in 10 large beaches. For example, 10 Copacabana Beaches have one exagrain of sand. Up again, we arrive at 10^{21}, one sextillion, or *zetta*. The number of stars in the observable universe is one zettastars. We run out of prefixes after *yotta*, which stands for 10^{24}, one septillion. But numbers can always get larger. The number 10^{100} is called a *googol*—yes, you probably know a company that has named itself after a purposeful misspelling. And then there is 10 raised to the googol power, $10^{10^{100}}$, which is one *googolplex*.[6]

These analogies will help us appreciate the relative merits of specific algorithms that we will examine in the rest of the book. Although in theory we could have algorithms of any kind of complexity, the algorithms we usually deal with fall into few different groups.

Complexity Families

The fastest family of all algorithms comprises the algorithms that run in no more than constant time, no matter what their input. We denote this complexity with $O(1)$; for example, an algorithm that checks if the last digit of a number is odd or even will not be affected by the size of the number and will run in constant time. The 1 in $O(1)$ follows from the fact that $O(1)$ means that the algorithm needs no more than a multiple of one steps to run—that is, a constant number of steps.

Before we meet the next complexity family, we need to take a brief excursion into a particular way things can grow or shrink. If you add something many times, you multiply it. If you multiply something many times, you raise it to a power or exponentiate it. We just saw how big numbers with exponents like 10^{12} (or more) can get. What is perhaps not immediately obvious is how quickly exponentiation leads to dizzying escalation—a phenomenon called *exponential growth*.

The probably apocryphal story about the invention of chess is illustrative. The ruler of the country where chess was invented asked its inventor what he would like for a gift (alas, it is a "he" in these stories). He replied that he would like one grain of rice on the first square of the chessboard, two on the second, four on the third, and so on. The king thought that he got off easily and granted the wish. Unfortunately, things quickly turned sour. The sequence grows in powers of two: $2^0 = 1$ in the first square, $2^1 = 2$ in the second square, $2^2 = 4$ in the third square, and thus in the last square the number of grains would be 2^{63}, a quantity unreachable by any means (it is equal to 9,223,372,036,854,775,808, or about 9 quintillion).

Exponential growth can also help us understand why it is so difficult to fold a piece of paper many times. Each time you fold it, you double the number of layers of the folded paper. After 10 folds, you have $2^{10} = 1{,}024$ layers. If your sheet is 0.1 millimeters thick, you now have a folded wad that is over 10 centimeters thick. Apart from the sheer force you will need to fold that in two, it may not be physically possible at all to do it, because to fold something it must be longer than thick.[7]

Exponential growth is the reason why computers have gotten more and more powerful over the years. According to *Moore's law*, the number of transistors in an integrated circuit doubles about every two years. The law is named after Gordon Moore, who founded Fairchild Semiconductor and Intel. He made the observation in 1965; the law proved prescient, so that we have gone from about 2,000 transistors in a processor in 1971 (the Intel

4004) to more than 19 billion in 2017 (the 32-core AMD Epyc).[8]

Having seen growth, let us explore now its opposite. If you have a multiple of something, you use division to reverse the operation and get the original value. If you have the power of something, a^n, how do you reverse the operation? The reverse of raising to a power is the *logarithm*.

Logarithms are sometimes taken as the boundary between mathematics for all and mathematics for the initiated; even the name has an aura of incomprehension. If logarithms seem somewhat hazy, you need to keep in mind that the logarithm of a number is the reverse of raising the number to a power. Just as when we raise to a power, we multiply repeatedly, when we take a logarithm, we divide repeatedly.

The logarithm is the answer to the question, "To which power should I raise a number to get the value I want?" The number we are raising is called the *base* of the logarithm. So if the question is, "To which power should I raise 10 to get 1000?," the answer is 3 because $10^3 = 1{,}000$. Of course, we may want to raise a different number—that is, use a different base. The notation for logarithms is $log_a x$ and it corresponds to the question, "To which power should I raise a to get x?" When $a = 10$, we just drop the subscript, because logarithms base 10 are common, so instead of writing $log_{10} x$ we simply write $log x$.

There are also two other common bases. When the base is the mathematical constant e, we write $\ln x$. The mathematical constant e, called *Euler's number*, is approximately equal to 2.71828. In the natural sciences we meet $\ln x$ a lot, which is why it is called *natural logarithm*. The other common base is 2, and instead of writing $\log_2 x$ we write $\lg x$. Base 2 logarithms are common in computer science and algorithms, but probably unused outside these fields, although we have already met them. In paper folding, if a wad of paper has 1,024 layers, it has been folded $\lg 1024 = \lg 2^{10} = 10$ times. In the chess example, the number of grains of rice results from the number of doublings we perform, which are $\lg 2^{63} = 63$.

The reason we see $\lg x$ a lot in algorithms is that it appears whenever we solve a problem by splitting it in two equal smaller problems; this is called *divide and conquer*, and it works like folding a sheet in two. The most efficient way to search for something in a sorted group of items has complexity $O(\lg n)$. That is pretty amazing; it entails that to find something among one billion ordered items, you need only $\lg 10^9 \approx 30$ probes into your items.

Algorithms that have logarithmic complexity are the next best thing after algorithms that run in constant time. Next come algorithms that run in $O(n)$, which are called *linear time* algorithms because their time grows proportionally with n; that means that they grow as multiples of n. We saw that searching for an item in an unordered set of items requires time proportional to the number of the items, $O(n)$. See how the complexity increased compared to when the items are ordered; organizing the data of our problem can have a big impact on how it can be solved. In general, linear time is the best behavior we can expect of an algorithm if it has to read through all the inputs of the problem, as this will require time $O(n)$ for n inputs.

If we combine linear and logarithmic times, we get *loglinear time* algorithms, where their time grows by n multiplied by its logarithm, $n\lg n$. The best algorithms for sorting—that is, putting items in order—have complexity $O(n\lg n)$. That may look a bit surprising; after all, it can be shown that if you have n items and want to compare each item with all other items, it requires time $O(n^2)$, which is bigger than $O(n\lg n)$.[9] Also, if you have n items that you want to sort, you definitely need $O(n)$ time to examine all of them. Sorting them requires multiplying that number by a smaller factor than n itself. We'll see how this can be done, later on in the book.

The next computational complexity family is n raised to a constant power, $O(n^k)$; this is called *polynomial complexity*. Polynomial time algorithms are efficient, except if k is big, but this rarely happens. When we try to solve a computational problem, we are usually delighted if we can come up with a polynomial time algorithm.

A complexity of the form $O(k^n)$ is called *exponential complexity*. Note the difference with the polynomial complexity where the exponent was constant; here it is the exponent that changes. We saw how exponential growth explodes. The universe will not survive long enough to see the answer of exponential algorithms for nontrivial inputs. Such algorithms are interesting from a theoretical point of view because they show that a solution can be found. We can then search for better algorithms with lower complexity, or we may be able to prove that no better algorithms can be found, in which case we can settle for something less than the ideal—for instance, approximate solutions.

There is something that grows even faster than exponentiation, and this is the *factorial*. If you have not encountered a factorial before, the factorial of a natural number n—which we write as $n!$—is simply the product of all the natural numbers up to and including that number: $100! = 1 \times 2 \times 3 \times \cdots \times 100$. Even if you have not encountered 100! you probably have encountered 52! even without knowing it. That is the number of different shuffles of a deck of cards. Algorithms whose running time is measured in factorials have *factorial complexity*.

Although numbers like 100! may seem exotic, they arise in many nonexotic settings and not just card games. Take, for example, the following problem: "If we have a list of cities and the distances between each pair of them, what is the shortest possible route that one should take to visit each city once and return to the origin city?" This is called the *traveling salesman problem*, and the obvious way to solve it is to examine every possible path taking in all cities. Unfortunately, for n cities this is n! The problem is unmanageable after, say, 20 cities. There are some algorithms that do it a bit better than $O(n!)$, but not enough to be practical. Surprising as it may seem for such a straightforward problem, the only way we can solve it in an acceptable time is by finding a solution that may not be the optimal one, but is close enough to it. Many other problems of great practical importance are *intractable*—that is, we don't know a practical algorithm to solve them exactly. Even so, the quest for better and better *approximation* algorithms is a vibrant field in computer science.

In the table that follows, you can see the value of various functions, falling under the complexity families we presented, for different values of n. The first row gives the n values and also stands in for linear complexity; subsequent rows show families of increasing complexity. As n increases, the function values increase, but the way they increase is different. The function n^3 will take us from one million to one quintillion, but that is nothing compared to 2^{100} or 100! We have left a blank like after the n^k row, separating practical from impractical algorithms. The border between the two are the polynomial algorithms, which as we saw are of practical use. Algorithms with higher complexity are usually not of practical use.

n	1	10	100	1,000	1,000,000
lgn	0	3.32	6.64	9.97	19.93
$nlgn$	0	33.22	664.39	9,965.78	1.9×10^7
n^2	1	100	10,000	1,000,000	10^{12}
n^3	1	1,000	1,000,000	10^9	10^{18}
n^k	1	10^k	100^k	$1,000^k$	$1,000,000^k$
2^n	2	1,024	1.3×10^{30}	10^{301}	$10^{10^{5.5}}$
$n!$	1	3,628,800	9.33×10^{157}	4×10^{2567}	$10^{10^{6.7}}$

2

GRAPHS

In the eighteenth century, the good citizens of Königsberg strolled around their city on Sunday afternoons. The city of Königsberg was built on the banks of the river Pregel. The river created two large islands within the city; the islands were connected to the mainland and each other with seven bridges in total.

Swept by the vagaries of European history, Königsberg passed from the Teutonic nights, to Prussia, Russia, the Weimer Republic, and Nazi Germany, and after the Second World War, it became part of the USSR and was renamed Kaliningrad, which is the name of the city today. It is part of Russia now, although not connected to Russia proper. Kaliningrad is situated in a Russian enclave, on the Baltic Sea, wedged between Poland and Lithuania.

Back in the day, the problem occupying the minds of the good citizens was whether it was possible to make their walks while crossing all seven bridges exactly once. The concern was named after its host city as the *Königsberg bridge problem*. To get a glimpse of the nature of the issue, here is a drawing of Königsberg at the time. The bridges are indicated by ovals drawn around them. The city had two islands, but you can see only one island in its entirety; the other one extends to the right beyond the boundaries of the map.[1]

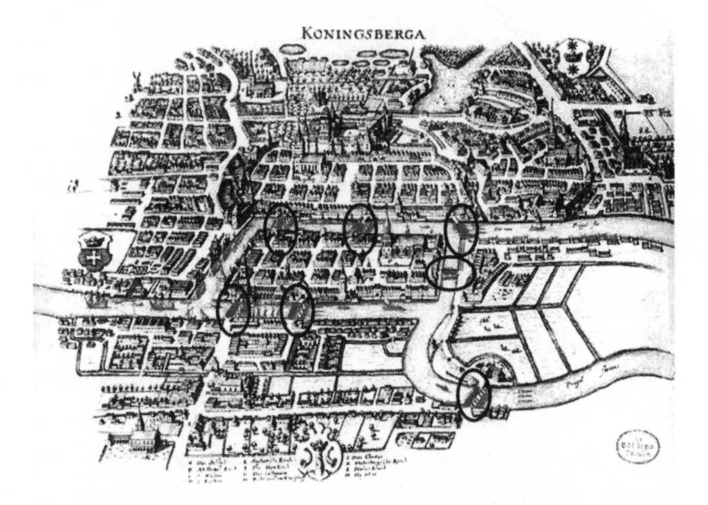

We don't know exactly how, but the famous Swiss mathematician Leonhard Euler learned about the problem; the problem is mentioned in a letter sent on March 9, 1736, from the mayor of Danzig, a city in Prussia 80 miles to the east of Königsberg (Danzig is now called Gdansk and belongs to Poland). The correspondence with Euler seems to have been part of an effort by the mayor to encourage the growth of mathematics in Prussia.

Euler was at the time living in Saint Petersburg in Russia. He worked on the problem and presented a solution to the members of the Saint Petersburg Academy of Sciences on August 26, 1735. In the following year, Euler wrote a paper, in Latin, describing his solution.[2] The solution was *negative*: it was not possible to make a tour of the city crossing each bridge only once. That would be an interesting piece of mathematical history, but by solving the problem, Euler created a whole new branch of mathematics: the study of *graphs*.[3]

Before we go into graphs, let's see how Euler tackled the problem. First of all, he abstracted the problem to its bare essentials. No detailed map of Königsberg is needed to represent the question. Euler drew the following diagram:[4]

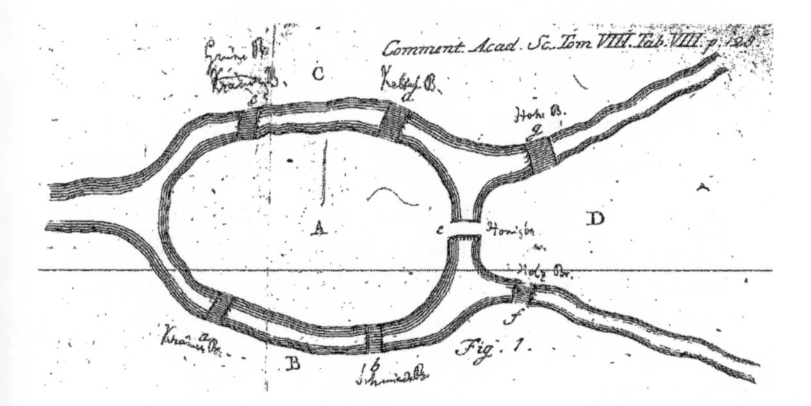

He used the letters A and D for the two islands, and B and C for the two banks on the mainland. The next step is to abstract the diagram even more, away from the physical geometry, and to the connections between bridges, islands, and mainland, because this is what really matters for the problem:

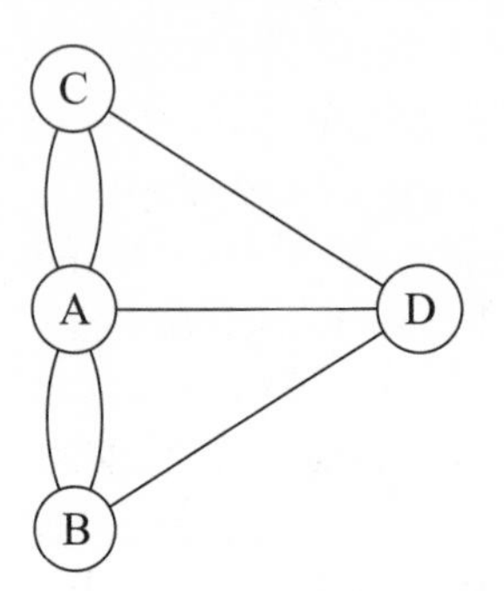

We have drawn the landmasses as circles, and the bridges as lines connecting the circles. The problem then can be restated as follows: If you have a pencil, is it possible to start from any of the circles, put the pencil down, and follow the lines without lifting the pencil from the paper so that you can pass through every line exactly once?

Euler's solution went as follows. Whenever you enter a landmass, you must leave it, except if this is the start or end of your walk. In order to do that, each landmass, apart from the start and finish, must have an even number of bridges so that each time you enter it, you can leave it from a different bridge, as required. Now go to the figure and count the number of bridges connecting each landmass. You will find out that all landmasses are connected with an odd number of bridges: A has five bridges, and B, C, and D have three bridges. Whichever of the landmasses we choose as starting and ending points, there will be two other landmasses that we will visit in the midst of our tour, and they have an odd number of bridges each. We cannot enter and leave them traversing their bridges only once.

Indeed, if we arrive at B at some point on our tour, we must have crossed a bridge to get to it. We will cross a second bridge to leave it. We must cross the third bridge at some later time because we are required to cross all bridges. But then we are stuck at B because there is no fourth bridge and we cannot cross a second time a bridge that we have already crossed. The same goes for C and D, which also have three bridges. Exactly the same argument holds for A as an intermediate point as it has five bridges; after crossing all five bridges of A, we won't be able to leave it from a different, sixth bridge because such a bridge does not exist.

The figure we drew consists of circles and lines connecting them. To use the proper terminology, we created a structure that is composed of *nodes* or *vertices* connected with *edges* or *links* between them. A structure that is composed of sets of nodes and edges is a *graph*; Euler was the first to recognize graphs as a structure and explore their properties. In today's language, the Königsberg bridge problem deals with *paths*: a path in a graph is a sequence of edges that connect a sequence of nodes. Then the Königsberg problem is the problem of finding a *Eulerian path* or *Eulerian walk*: a trail through a graph such that each edge is visited exactly once. A path that starts and ends at the same node is called a *tour* or *circuit*. If we also add the restriction (not in the original problem) that we want the Eulerian path to start and finish at the same point, then we have a *Eulerian tour* or *Eulerian circuit*.

The applications of graphs are so numerous that they fill entire books. Anything that can be modeled by nodes connected to other nodes can be represented as a graph. Once we do that, we can ask all kinds of interesting questions about it; here we'll have the opportunity to take just a glance.

Before we do that, though, here is a small detail to please the most rigorous minded of readers. We mentioned that a graph is a structure that comprises sets of vertices and edges. In mathematics, a set does not contain the same item twice. Yet in our representation of Königsberg, we have the same edge appear more than once; there are, for example, two edges between A and B. An edge is distinguished by its starting and ending points, so the two edges between A and B are in fact two instances of the same edge. Then the set of the edges is not really a set; it is a *multiset*—that is, a set that allows for multiple instances of its elements. In the same way, the Königsberg graph is not really a graph but rather a *multigraph*.

From Graphs to Algorithms

The definition of a graph is wide so that it can encompass everything that can be represented as things connected to other things. The graph may have some relevance to the topology of a place, but the nodes and links may have nothing to do with points in space.

A *social network* is an example of such a graph. In a social network, nodes are social actors (these may be individuals or organizations), and the links represent interactions between them. The social actors may be real-world actors, and the links may be their collaborations in films. The social actors can be us, and the links may be our connections to other people in a social network application. We can then use social networks to find communities of people, starting from the premise that communities are formed by people who interact with each other. There exist algorithms that are able to find efficiently communities in graphs with millions of nodes.

The edges in the Königsberg graph are not directed, meaning that we can traverse them both ways; for example, we can go from A to B and B to A. The same goes for

The definition of a graph is wide so that it can encompass everything that can be represented as things connected to other things.

social networks, when the connections are reciprocal. That is not always necessary. Depending on our applications, edges in a graph may be directed. When this happens, we draw the edges with arrows at their ends. Directed graphs are called *digraphs* for short. You can see a digraph below. Note that this is not a multigraph; the edge from A to B is not the same as the edge from B to A.

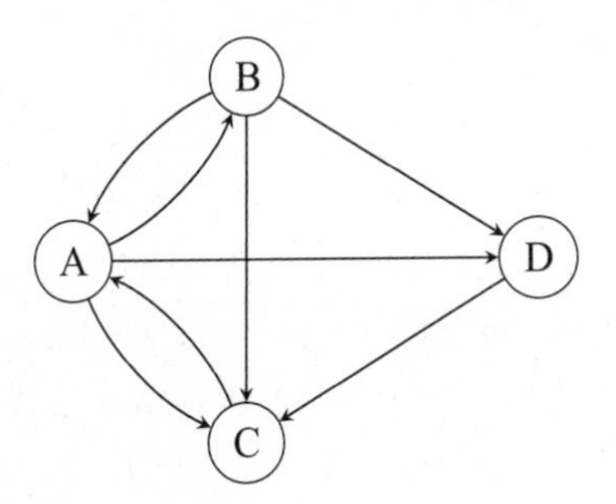

The World Wide Web is an example of a (huge) directed graph. We can represent the web with nodes standing in for web pages and edges standing in for the hyperlinks between each pair of pages. This graph is a directed graph, because a page may link to another page, but that other page does not necessarily link back to the first page.

When it is possible to start from a node in a graph, traverse edges, and come back to the node we started from, we say that the graph has a *cycle*. Not all graphs have cycles. The Königsberg graph has cycles—although it does not have a Eulerian circuit. A famous cyclic graph (actually a multigraph) in the history of science is August Kekulé's model of the molecular structure of benzene:[5]

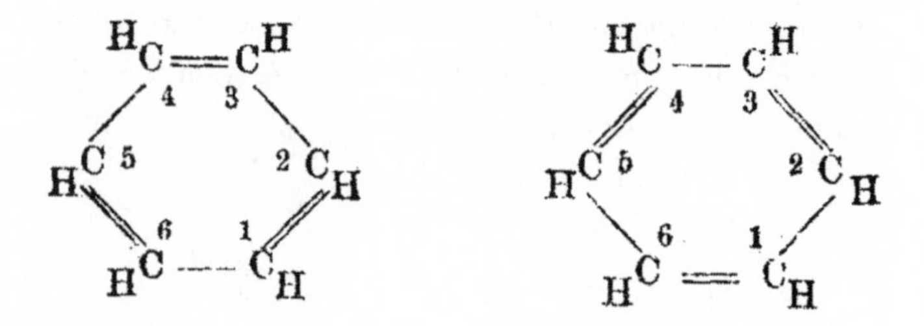

A graph without a cycle is called an *acyclic graph*. Directed acyclic graphs form an important class of graphs. We usually call them *dags*. Dags have many uses; for example, we use them to represent priorities between tasks (tasks are nodes, and priorities are links between them), dependency relations, prerequisites, and other similar arrangements. We'll leave aside acyclic graphs now and turn our attention to cyclic graphs, which will provide us with a first window on algorithms on graphs.

Paths and DNA

One of the most important scientific developments of the last decades has been the decoding of the human genome.

Thanks to the techniques that were developed in that effort, we can now investigate genetic diseases, detect mutations, and study genomes of extinct species, among other fascinating applications.

Genomes are encoded in the DNA, a large organic molecule that is composed of a double helix. The double helix is made up of four bases: cytosine (C), guanine (G), adenine (A), and thymine (T). Each part of the double helix is constructed from a series of bases, like ACCGTATAG. The other part of the double helix is constructed from bases that are connected with their corresponding bases on the first part, according to the rules A-T and C-G. So if one part of the helix is ACCGTATAG, the other part will be TGGCATATC.

In order to find the composition of an unknown DNA piece, we can work as follows. We create many copies of the chain and break them up into little fragments—for instance, fragments containing three bases each. Using specialized instruments, we can identify such small fragments easily. In this way we end up with a set of known fragments. We are then left with the problem of assembling the fragments into a DNA sequence, whose composition we will then know.

Suppose then that we have the following fragments, or polymers as they are known: GTG, TGG, ATG, GGC, GCG, CGT, GCA, TGC, CAA, and AAT. Each one of them has a length of three; to find the DNA sequence from which they were broken up, we create a graph. In that graph, the vertices are polymers of length two that are derived from the polymers of length three, taking for each polymer of length three the first two and last two polymers. So from GTG we will get GT and TG, and from TGG we will get TG and GG. In the graph, we add one edge for every one of the initial polymers or length three that was used to derive the two vertices. We give the name of the polymer to that edge. From ATG we get vertices AT and TG and the edge ATG. You can see the graph that results from our example:

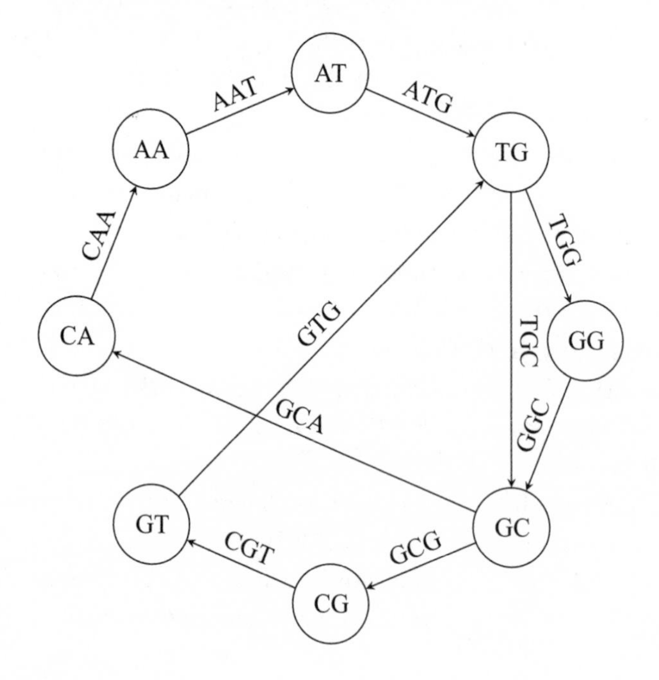

With the graph we have created, we only need to find a tour in the graph that visits all edges exactly once—that is, an Eulerian circuit—in order to find the initial DNA sequence. The *Hierholzer algorithm* for finding Eulerian circuits on graphs was published by the German mathematician Carl Hierholzer in 1873 and goes like this:[6]

1. We pick a starting node.

2. We go from node to node until we return to the starting node. The tour that we have traced to this point does not necessarily cover all edges.

3. As long as there exists a vertex that belongs to the tour we have traced, but is also part of an edge that is not in the path, we start another path from that vertex, using edges that we have not used yet, until we return to it, forming another tour. Then we splice this tour to the tour we have already traced.

If we use the algorithm in our example graph, we will find the path in the following figure:

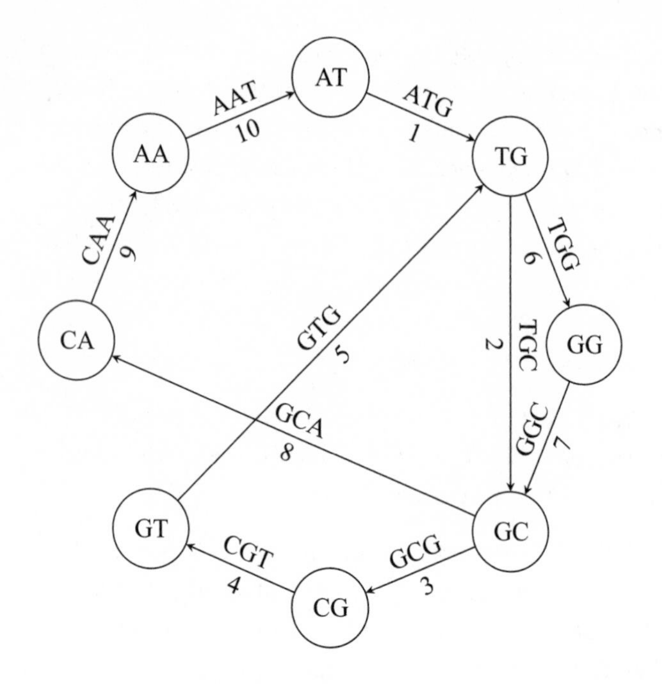

We started from AT and made the tour AT → TG → GG → GC → CA → AA → AT. We made a tour, but we did not cover all the edges. We see that TG has an edge, TGC, that we have not covered yet. So we go to TG and do a tour starting along the TGC edge, getting TG → GC → CG → GT → TG. We splice the second path into the first, getting the one in the figure, AT → TG (→ GC → CG → GT → TG) → GG → GC → CA → AA → AT. If we walk the resulting path from the first node to the last, without stepping on the last, and concatenate the vertices keeping their common base only once, we get the DNA sequence ATGCGTGGCA. You can verify that this sequence contains all the polymers with which we started; CAA and CAT are found if you wrap around when you reach the end of the sequence and go to the beginning.

In this particular illustration, we only found one additional tour that we spliced into the original one. In general, there may be more; step 3 of the algorithm is repeated as long as there are vertices with edges that we have not covered yet. Hierholzer's algorithm is fast: if implemented properly, it runs in linear time, $O(n)$, where n is the number of edges in the graph.[7]

Scheduling a Tournament

Suppose you are organizing a tournament in which the contestants will compete in pairs, so we'll have a series of matches. We have eight contestants, and each contestant will play four matches. Our problem is how to schedule the tournament. We want to schedule the matches so that each contestant plays only one match per day.

An obvious solution is to have just one match per day and allow the tournament to last as long as needed. As we have eight contestants and each contestant plays four matches, the tournament would roll out over 16 days

($8 \times 4 / 2$; we divide by two so as not to count each match twice). We'll name the eight contestants Alice, Bob, Carol, Dave, Eve, Frank, Grace, and Heidi. This allows us to use only the initial letter of their names to identify them.

We can find a better solution if we model the problem as a graph. We'll have a vertex for each player and an edge for each match. Then the graph will look like the one on the left below. On the right, we have labeled the edges with the day on which the corresponding match will take place. How did we find this solution?

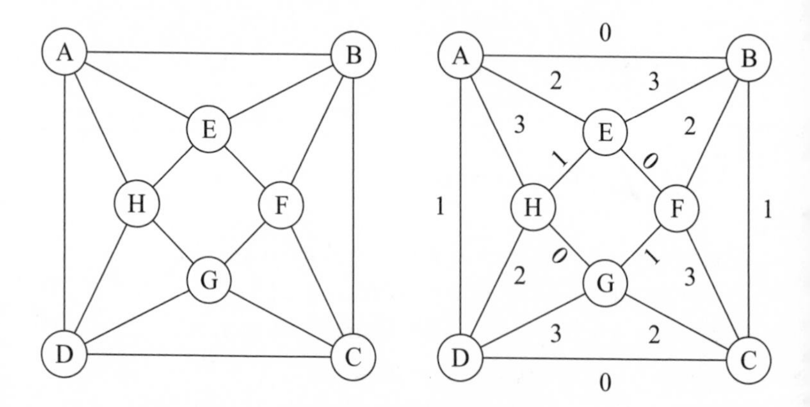

We agree to number the tournament days consecutively. Let the tournament start on day zero. We'll schedule all matches, one by one.

1. Take a match that we have not scheduled yet. If we have scheduled all matches, stop.

2. Schedule the match on the earliest day so that neither of the two players has another match on that day. Return to step 1.

This algorithm looks deceptively simple, and you may doubt that it really solves our problem. So let's walk through it and see what happens. In the following table we can see the matches, one by one, and the day on which we schedule each match, as we apply the algorithm on the graph. You should read the first two columns of the table and then the next two:

Match	Day	Match	Day
A, B	0	C, F	3
A, D	1	C, G	2
A, E	2	D, G	3
A, H	3	D, H	2
B, C	1	E, F	0
B, E	3	E, H	1
B, F	2	F, G	1
C, D	0	G, H	0

We start by taking the match Alice versus Bob. Neither Alice nor Bob play any other match on day zero—that is, the day on which we'll assign the match.

We then take another match we have not scheduled yet—say, Alice versus Dave. Although there is no requirement to do so, we'll take the match players in lexicographical order as we continue, but bear in mind that we could take them in any other way, even randomly, as long as we treat each match only once. Alice already has a match scheduled on day zero, so the earliest available day for the match is day one.

Next comes the match between Alice and Eve. Alice is booked on day zero and day one, so we'll schedule it on day two. Alice's final match will be with Heidi; Alice is engaged on days zero, one, and two, so this will have to go on day three.

We are done with Alice. Moving on to Bob's matches, except for the one with Alice, which we have already scheduled, we need to plan Bob versus Carol. Bob is already scheduled on day zero (with Alice), so this match will have to go on day one. Scheduling Bob versus Eve, we notice that Bob is already engaged on day zero and day one (we just scheduled that), while Eve is scheduled to play on day two with Alice; we therefore schedule Bob versus Eve on day three. Going to Bob versus Frank, Bob has matches on days zero and one, but is free on day two, while Frank has no matches at all as of yet. So Bob versus Frank goes on day *two*, earlier than Bob versus Eve.

After Bob, we'll deal with Carol's matches. Neither Carol nor Dave have a match scheduled on day zero, so Carol versus Dave will go on the first day of the tournament. After this, the Carol versus Frank match can take place on day three, because Carol plays matches on day zero (we just arranged that) and day one (with Bob, arranged previously), while Frank plays with Bob on day two (also arranged previously). Carol versus Grace will take place *earlier*, on day two, as Grace has no other matches planned as of yet and Carol is still free on day two.

We proceed similarly with the rest of the matches; it is interesting that the matches in the inner and outer squares of the graph will happen as early as the first two days. These are two different groups playing in parallel before they start playing between them. At the end, the solution we find is a significant improvement over the naive solution requiring 16 days; we only need four!

This tournament scheduling problem is in fact an instance of a more general problem: the *edge coloring* problem. An edge coloring of the graph is an assignment of colors to edges so that no two adjacent edges have the same color. Now color should be taken figuratively here. In our example, the colors are the days; in general, they can be any other set of distinct values. If instead of the edges, we want to color the vertices of the graph so that no two vertices that are linked by an edge share the same color, then we have the *vertex coloring* problem. Both edge and vertex coloring belong to the wider class of, no surprise, *graph coloring* problems.

The algorithm we described for edge coloring is simple and efficient (it takes each edge one by one, and only once). It is a so-called *greedy algorithm*. Greedy algorithms are algorithms that try to solve a problem by finding the best solution *at each stage*, not the optimal solution in general. Greedy algorithms are useful in many problems when at each stage of the solution we have a choice to make and our rule is "what looks best now." Such strategies that guide our choices in the evolution of an algorithm are called *heuristics*, from the Greek *heuriskein*, which means "to find" (a solution, that is).

With some thought we can realize that in algorithms, as in real life, what looks best right now may not really be the best strategy. It may pay off to delay gratification; the best choice right now may lead us to a trap that we'll regret later on. Imagine you are climbing a mountain. The greedy heuristic would be to select the steepest path at each point (we assume that your climbing prowess is unparalleled). This will not necessary lead you to the top: it may well lead you to a plateau, from which the only way is back. The real way to the top may lie through gentler slopes.

The climbing metaphor is frequently used in problem solving in computer science. We model our problem so that the solution lies at "the top" of the possible moves we can make and try to find the correct moves; this is called a *hill climbing* approach. When we arrive at something like a plateau, we say we arrived at a *local optimum*, but not the *global optimum*, the highest peak that we are after.

From hill climbing back to tournament scheduling, we selected the first available day for each match. Unfortunately, this might not be the best way to schedule all matches. Indeed, it turns out that graph coloring is a difficult problem. The algorithm that we gave is *not* guaranteed to give the optimal solution—that is, the solution requiring the smallest number of days (or colors, in general). The number of edges adjacent to a node is called its *degree*. It can be proven that if the largest degree of any node in the graph is d, the edges can be colored with at most d or $d + 1$ colors; the required number of colors for the edges of a graph is called its *chromatic index*. In our particular example, the solution is optimal, $d = 4$, and we used four days. Our algorithm, however, may not be able to find the optimal solution in some other graph. It may give us a solution worse than that. The good thing about greedy graph coloring is that we know how far off that solution might be: the solution it will give may need up to $2d - 1$ colors, instead of d, but no worse than that.

If you want to see how this may happen, consider a graph that consists of "stars" connected to a central node, like the one on the next page:

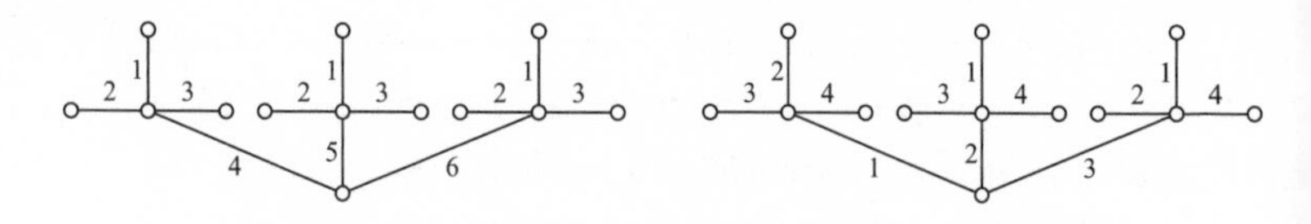

If we have k stars, where each star has k edges plus an edge to the central node, and we start by coloring the stars, we'll use k colors to color the edges of the stars. Then we'll need k additional colors to connect the stars to the central node. The total is $2k$ colors. This is what we did on the left. But this is not the optimal solution. If we start by coloring the edges connecting the stars to the central node, we'll need k colors for that. Then we can color the stars themselves using only one additional color, for a total of $k + 1$ colors. You can see how we can do that on the right. All this is in accordance with theory, as each star has degree $k + 1$.

The problem is that the greedy algorithm decides to order the edges to color in a way that is not optimal at the end, or to use the proper terminology, in a way that is not *globally optimal*. It might hit on the best solution, but it might not. Then again, the difference from the optimum solution is not that great. That is a relief because graph coloring is so difficult that if we want an exact algorithm that can find the best solution for every graph, the algorithm will have exponential complexity, about $O(2^n)$, where n is the number of edges in the graph. Exact edge coloring algorithms are therefore useless, except for tiny graphs.

The greedy algorithm we have presented has one additional nice property (apart from being practical). It is an *online algorithm*: an algorithm that works even if the inputs are not known when we start but instead arrive on the scene as we go. We don't need to know all the edges to start running the algorithm. The algorithm will work correctly, even if the graph is constructed in a piecemeal fashion, one edge at a time, while we are running the algorithm. This would happen if players are signing up for the tournament even after we have started scheduling the matches. We will be able to color each edge (match) as it comes, and whenever the graph is finished, we'll have an edge coloring ready. Moreover, this greedy algorithm is the optimum algorithm if the graph is created incrementally in this way; no exact algorithm, no matter how inefficient, exists at all when the graph is constructed while we are solving the problem.[8]

Shortest Paths

As we saw, a greedy algorithm works by taking the best decision at each step—which may not be the best decision overall. It has a somehow opportunistic nature or carpe diem feeling to it. Unfortunately, as Aesop's fable tells us, a grasshopper living for the day may yet live to regret the winter, when the ant, who is preparing for the future, ends up cozy and warm.[9] In the planning of tournaments, we found that the grasshopper may not end up so badly. Now it is time for the ant's revenge.

In chapter 1, we discussed the infeasibility of trying to find the shortest path between two points on a grid by enumerating all the possible paths. We saw that this is impossible to do in practice because the number of paths increases tremendously. Now with our knowledge of graphs, we will see that there is a way. In fact, we'll take the problem up a notch. Instead of looking for the shortest path on a grid, which has a kind of nice geometry and on which all distances between points are equal, we will allow any geometric shape and even add different distances between points.

To do that, we'll create a graph where we have nodes and edges representing a map, and want to find the shortest way between two nodes on the map. Moreover, we'll attach a *weight* to each edge. The weight may be positive or zero, and will correspond to a measure of the distance between the two connected nodes. It may be distance in miles or travel time in hours; any other nonnegative metric will do. Then the *path length* is the sum of the weights along the path; the *shortest path* between two nodes is the path with the smallest length. If all weights are equal to one, then the path length is equal to the number of edges on the path. Once we allow weights to have other values, this is no longer true.

In the following graph, we have six nodes connected by nine edges with varying weights, and want to find the shortest path to travel from nodes A to F.

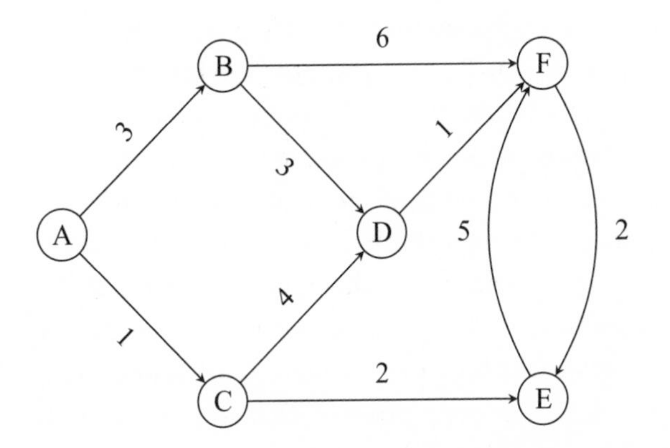

If we adopt a greedy heuristic, we'll start by going from node A to C, then the best choice is to go to node E, and from there we make our way to node F. The total length of the path A, C, E, and F is eight, which is not, however, the best path. The best path is to go from A to C to D, and then to F, for a total length of six. So the greedy heuristic does not work, and in contrast to tournament planning, there are no guarantees as to its worst performance in relation to the actual shortest path. Nevertheless, and again in contrast to tournament planning, there exist efficient algorithms for finding the shortest paths so in fact there is no reason to use the greedy heuristic at all.

In 1956, a young Dutch computer scientist, Edsger Dijkstra, was shopping in Amsterdam with his fiancée. Having got tired, they sat down at a café terrace to drink a cup of coffee, where Dijkstra thought about the problem of finding the best way to go from one city to another. He designed the solution in 20 minutes, although the algorithm took some time, three years, to get published. Dijkstra led an illustrious career, yet this 20-minute invention remained, to his amazement, a cornerstone of his fame.[10]

So how does the algorithm go? We want to find the shortest paths from one node to all other nodes in a graph. The algorithm uses an idea called *relaxation*: we assign estimates for the values we want to find (here, distances). In the beginning, our estimates are the worst possible. Then as the algorithm progresses, we are able to relax these estimates from the extremely bad ones we started with to progressively better and better ones, until we arrive at the correct values.

In Dijkstra's algorithm, relaxation proceeds as follows. We begin by assigning the worst possible value for the distances of all nodes from our starting node: we set the distance to infinity; clearly there cannot be anything worse than that! In the following figure, we have placed the initial estimate for the shortest path and previous node in that path above or below each node. For the A node, we have 0 / – because the distance from A to A is zero and there is no previous node to A. For all other nodes, we have ∞ / – because the distance is infinity and we have no idea about the shortest path to them.

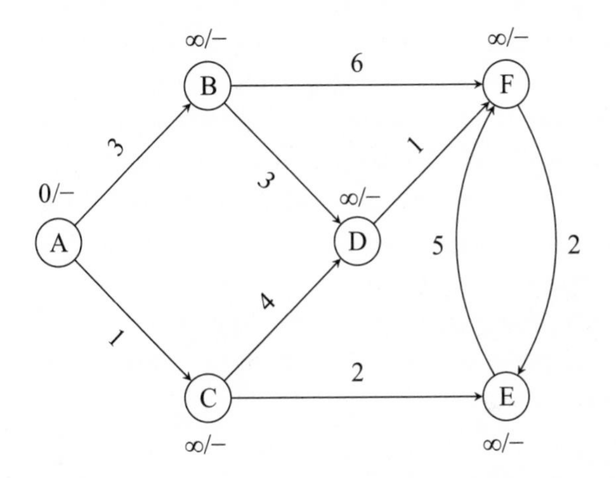

We take the node with the shortest distance from A thus far. This is A itself. That is our current node, so we mark it gray.

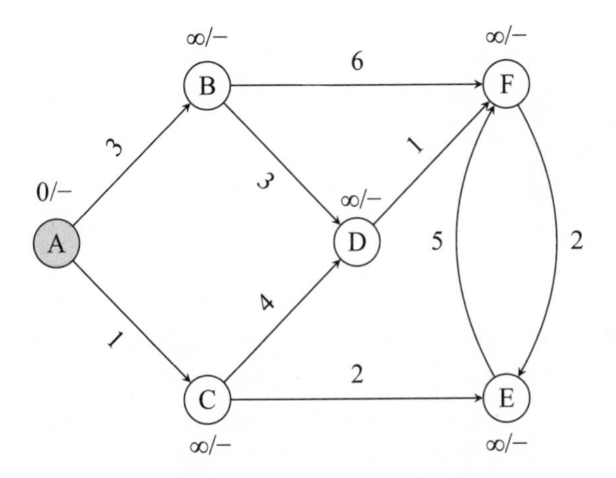

From A we can check the estimates for the shortest paths to its neighbors, B and C. Initially we had set them at infinity, but in fact now we find out that we can get to B from A at a cost of 3 and we can get to C from A at a cost of 1. We update these estimates and also indicate that the estimates are through A; we write 3/A above B and 1/A below C. We are done with node A for the rest of the algorithm. We update the figure accordingly, marking A black. We move to the unvisited node with the best current estimate. That is node C.

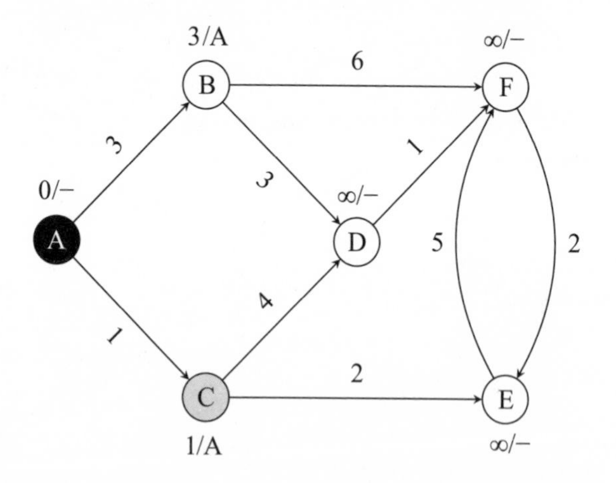

From node C, we check the estimates of the shortest paths to its neighbors, D and E. They were at infinity, but now we see that we can get to each one of them through C. The path from A to D through C has a total length of 5, so we write 5/C above D. The path from A to E through C has a total length of 3, so we write 3/C below E. We are done with node C so we mark it black and move to the unvisited node with the best current estimate. Both nodes B and E have an equally good estimate of 3. We can pick either. Let us pick B.

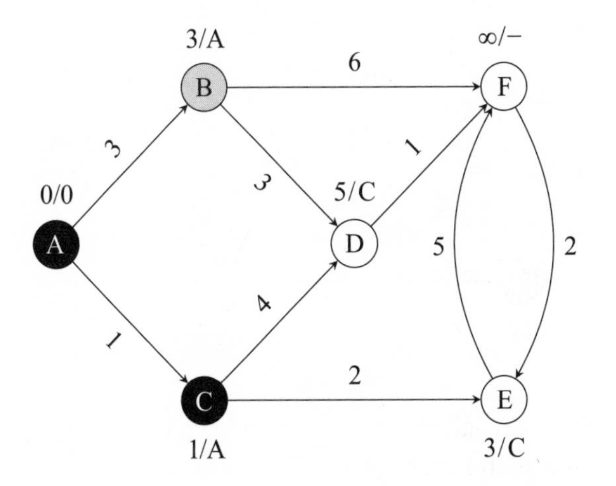

We work in the same way. From node B, we check the estimates of the shortest paths to its neighbors, D and F. We already have an estimate of length 5 for D, coming from C; that is better than the length 6 that we would get coming from B. So we let the estimate to D remain unchanged. The current estimate to F is infinite so we update it to 9, coming from B. We mark B as visited and move to the unvisited node with the best current estimate. That is node E.

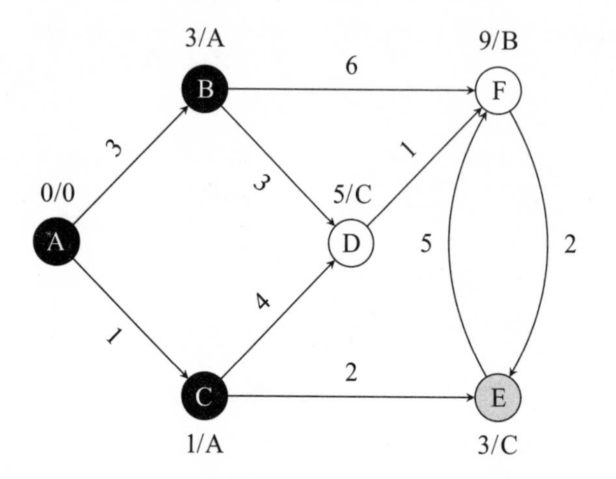

E has F as a neighbor. The path to F from E has length of 8, which is better than the path we had found through B. We update the path, mark E as visited, and move to the unvisited node with the best current estimate, node D.

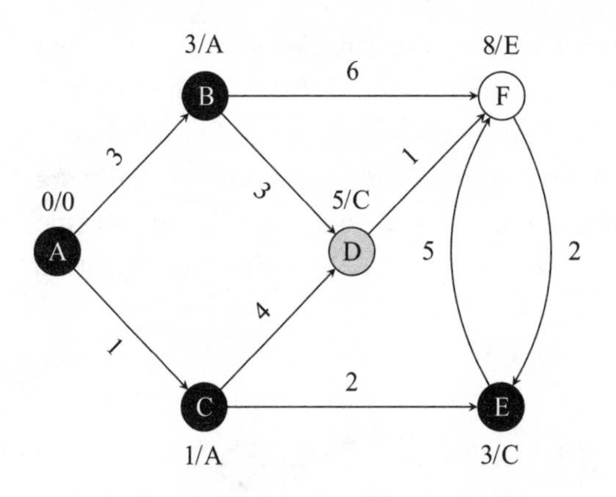

D has F as a neighbor, to which we have found a path coming from E with length 8. As we can get to F through D with a total length of 6, we update that path. As before, we move to the unvisited node with the best current estimate—actually our only unvisited node, F.

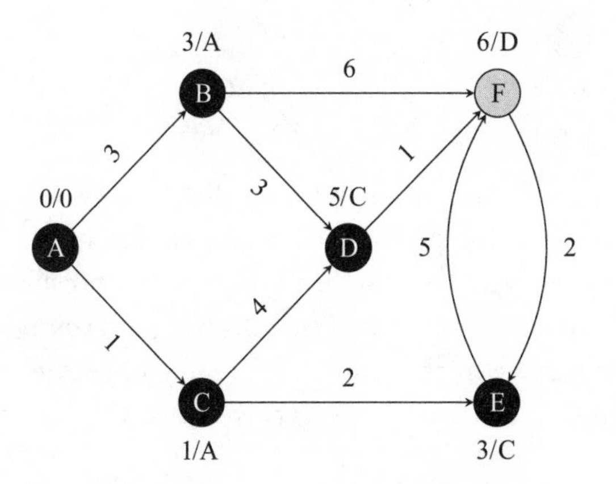

From node F we check whether we should update our estimate for its neighbor, node E. The current path to E has a length of 3, while the path through F would have a cost of 10. We let E remain unchanged. Visiting F did not make any difference, but we could not have known that beforehand. As we have visited all nodes, the algorithm finishes.

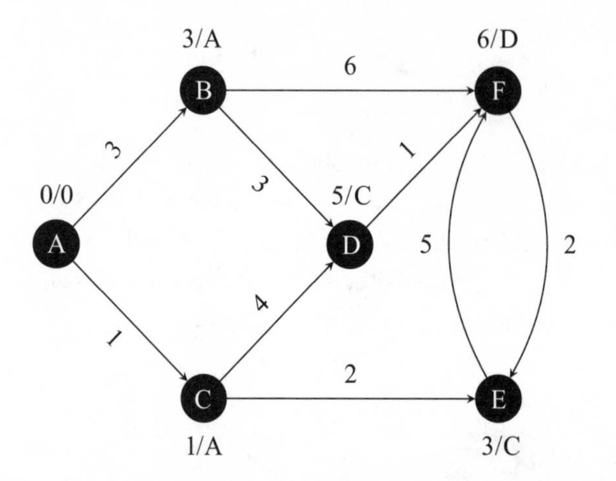

When we were going through the algorithm, we were recording path lengths and the predecessor of each node along the shortest path. We did that so that if after finishing the algorithm we want to find the shortest path from A to any other node in the graph—for example, F—we start from the end and make our way to the start. We read its predecessor: D. We get the predecessor of D, which is C, and then the predecessor of C, which is A. The shortest path from A to F is A, C, D, and F with a total length of six, as we had mentioned way back at the start of our discussion.

At the end, Dijkstra's algorithm found *all the shortest paths* from the starting node to all other nodes in the graph. The algorithm is efficient, as its complexity is $O((m + n)logn)$, where m is the number of edges in the graph and n is the number of nodes. Here is the algorithm as a set of steps:

1. Assign a distance equal to infinity to all nodes except for the starting node; assign a distance equal to zero for the starting node.

2. Find the unvisited node with the minimum distance. This will be our current node. If there is no unvisited node, stop.

3. Examine all neighbors of the current node. If their distance is greater than the distance we would get passing from the current node before arriving at the neighbor, we relax the distance and update the path going to the neighbor. Go to step 2.

If we are only interested in the shortest path to a particular node, we can stop when we pick it to visit in step 2.

Once we do that, we have already found the shortest path to it, and it will not change in the rest of the algorithm's execution.

We can use Dijkstra's algorithm in any graph, directed or not, even if it contains cycles, provided that it does not have negative weights. This might happen if the edges represent some kind of rewards and penalties between nodes. The good news is that there are other efficient algorithms that we can use in the presence of negative weights, but this highlights that algorithms may have particular requirements in their applicability. When we try to find an algorithm to solve our problem, we should check that our problem meets the requirements of the algorithm. Otherwise the algorithm will not work; but note that an algorithm cannot tell us that it does not work. If we implement the algorithm on a computer, it will still execute its steps even if it does not make sense to do so. It will produce an answer that will be nonsense. It is up to us to make sure that we are using the right tool for the right job.

For an extreme example, think of what would happen with a graph that not only has negative weights but also a cycle where the sum of the edges is negative: a negative cycle. Then *no algorithm* would find the shortest paths in the graph *because they do not exist*. If we have a negative cycle, we can go round and round its edges, and every time the length of the path will be reduced. We can continue forever, and the path along the cycle will get to negative

When we try to find an algorithm to solve our problem, we should check that our problem meets the requirements of the algorithm. Otherwise it will not work; but an algorithm cannot tell us that it does not work.

infinity. Computer scientists and programmers have a name for when we put something in a program that does not make sense for it: *garbage in, garbage out*. It is up to humans to ferret out the garbage and know what to use when. An important part of algorithm courses in universities is exactly to teach budding computer scientists what to use when.

3

SEARCHING

The fact that algorithms can do all sorts of stuff, from translating text to driving cars, can give us a misleading picture of what algorithms are mostly used for. The answer may seem mundane. It is unlikely that you will be able to find any computer program doing anything at all useful without employing algorithms for searching in data.

That is because searching in one form or another appears in almost every context. Programs take in data; often they will need to search for something in them and so a searching algorithm will almost certainly be used. Not only is searching a frequent operation in programs but, because it happens frequently, searching can be the most time-consuming operation in an application. A good search algorithm can result in dramatic improvements in speed.

A search involves looking for a particular item among a group of items. This general problem description

Searching in one form or another appears in almost every context. . . . A good search algorithm can result in dramatic improvements in speed.

encompasses several variations. It makes a big difference whether the items are ordered in some way that is related to our search or come in random order. A different scenario occurs when the items are given to us one by one and we have to decide if we have found the correct one right when we confront it, without the ability to rethink our decision. If we search repeatedly in a set of items, it is important to know if some items are more popular than others so that we end up searching for them more often. We will examine all these variations in this chapter, but keep in mind that there are more. For example, we will only present *exact search* problems, but there are many applications in which we need an *approximate search*. Think of spellchecking: when you mistype something, the spellchecker will have to search for words that are similar to the one it fails to recognize.

As the data volumes increase, the ability to search efficiently in a huge number of items has become more and more significant. We'll see that if our items are ordered, the search can scale extremely well. In chapter 1 we stated that it is possible to find something among a billion sorted items in about 30 probes; now we will see how this can be actually done.

Finally, a search algorithm will give us a glimpse of the dangers that lurk when we move from an algorithm to an actual implementation in a computer program, which has to run within the confines of a particular machine.

A Needle in a Haystack

The simplest way to search is what we do to find the proverbial needle in a haystack. If we want to find something in a group of objects and there is absolutely no structure in them, then the only thing we can do is to check one item after the other until we either find the item we are looking for or fail to find it after exhausting all items.

If you have a deck of cards and are looking for a particular one in them, you can start taking off the cards from the top of the deck until you find the one you are looking for or run out of cards. Alternatively, you can start taking off the cards one by one from the bottom of the deck. You can even take off cards from random positions in the deck. The principle is the same.

Usually we do not deal with physical objects in computers but rather digital representations of them. A common way to represent groups of data on a computer is in the form of a *list*. A list is a data structure that contains a group of things in such a way that from one item we can find the next one. We can usually think of the list as containing *linked items*, where one item points to the next one, until the end, where the last item points to nothing. The metaphor is not far from the truth because internally the computer uses memory locations to store items. In a *linked list*, each item contains two things: its payload data and the memory location of the next item on the list. A place in memory that holds the memory location of another place in memory is called a *pointer*. Therefore in a linked list, each element contains a pointer to the next element. The first item of a list is called its *head*. The items in a list are also called *nodes*. The last node does not point to anywhere; we say that it points to *null*: nothingness on a computer.

A list is a sequence of items, but it is not necessary that the sequence is ordered using some specific criterion. For example, the following is a list containing some letters from the alphabet:

U → R → L → A → E → K → D → ⊠

If we have an unordered list, the algorithm for finding an item on it goes like this:

1. Go to the head of the list.

2. If the item is the one we are looking for, report that it is found and stop.

3. Go to the next item on the list.

4. If we are at null, report that the search item was not found and stop. Otherwise, return to step 2.

This is called a *linear* or *sequential search*. There is nothing special about it; it is a straightforward implementation of the idea of examining each single thing in turn until we find the one we want. In reality, the algorithm makes the computer jump from pointer to pointer until it either reaches the item we are looking for or null. Below we show what is happening when we search for E or X:

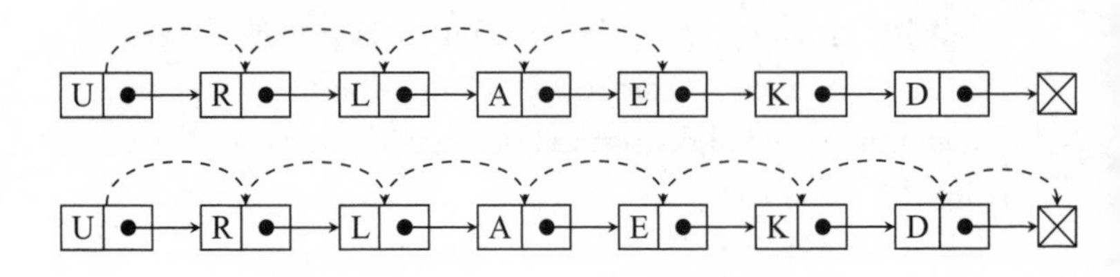

If we search among n items, the best thing that can happen is to hit on the item we want immediately, which will occur if it is the head of the list. The worst thing that can happen is that the item is the last one on the list or not on the list at all. Then we must go through all n items. Therefore the performance of sequential search is $O(n)$.

There is nothing we can do to improve on that time if the items appear on the list in a random sequence. Going back to a deck of cards, you can see why this is so: if the deck is properly shuffled, there is no way to know in advance where we'll find our card.

Sometimes people have trouble with that. If we are looking for a paper among a large pile, we may tire of going one after the other. We may even think of how unlucky we would be should the paper turn out to be at the bottom of the pile! So we stop going through the pile in order and peek at the bottom. There is nothing wrong in peeking at the bottom, but it's wrong to think that this improves our chances of finishing the search quickly. If the pile is random, then there is no reason why the sought-after item is not the first, last, or one right in the middle. Any position is equally likely, so starting from the top and making our way to the bottom of the pile is as good a strategy as any other that ensures we examine each item exactly once. It is usually simpler to keep track of what we looked at if we work in a specific order, however, than jumping around erratically, and that's why we prefer to stick with a sequential search.

All this holds as long as there is no reason to suspect that the search item is in a particular position. But if this is not true, then things change, and we can take advantage of any extra information we may have to speed up our search.

The Matthew Effect and Search

You may have noticed that in an untidy desk, some things find their way to the top of the pile, while some others seem to slip to the bottom. When finally cleaning up the mess, the author has had the pleasant experience of discovering buried deep down in a heap things he believed were long lost. The experience has probably occurred to others as well. We tend to place things we use frequently close; things we have little use for slip further and further out of reach.

Suppose we have a pile of documents on which we need to work. The documents are not ordered in any way. We go through the pile, searching for the document we need, processing it, and then placing it not where we found it but instead on the top of the pile. Then we go again with our business.

It may happen that we do not work with the same frequency on all documents. We may return to some of them again and again, while we may only rarely visit others. If we continue placing every document on the top of the pile after working on it, after some time we'll find out that the most popular documents will be near the top, while the ones we accessed the least often will have moved toward the bottom. This is convenient for us because we spend less time locating the frequently used documents and thus less time overall.

This suggests a general searching strategy, where we search for the same items repeatedly, and some items are more popular than others. After finding an item, bring it forward so that we'll be able to find it faster the next time we will look for it.

How applicable would such a strategy be? It depends on how often we observe such differences in popularity. It turns out that they happen a lot. We know the saying "the rich get richer, and the poor get poorer." It is not just about rich and poor people. The same thing appears to a bewildering array of aspects in different fields of activity. The phenomenon has a name, the *Matthew effect*, after the following verse in the Gospel of Matthew (25:29): "For unto every one that hath shall be given, and he shall have abundance: but from him that hath not shall be taken away even that which he hath."

The verse talks about material goods, so let's think about wealth for a minute. Suppose you have a large stadium, capable of holding 80,000 people. You are able to measure the average height of the people in the stadium. Your result may be something around 1.70 meters (5 feet, 7 inches). Imagine that you take out somebody randomly from the stadium and put in the tallest person in the world. Will the average height differ? Even if the tallest person is 3 meters tall (no such height has ever been recorded), the average height would remain stuck at its previous value—the difference with the previous average being less than a tenth of a millimeter.

Imagine now that instead of measuring the average height, you measure the average wealth. The average wealth of your 80,000 people could be $1 million (we are assuming a wealthy cohort). Now you substitute again somebody inside with the richest person in the world. That person could have a wealth of $100 billion. Would this make a difference? Yes, it would—and a big one. The average would increase from $1 million to $2,249,987.5, or more than double. We are aware that wealth is not distributed equally around the world, but we may not be aware of how unequal the distribution is. It is much more unequal than a distribution of natural measures like height.

The same difference in endowments occurs in many other settings. There are many actors you have never heard of. And there are a few stars who have appeared in many movies, earning millions of dollars. The term "Matthew effect" was coined by the sociologist Robert K. Merton in 1968, when he observed that famous scientists get more credit for their work over their lesser-known colleagues, even if their contributions are similar. The more famous scientists are, the more famous they will get.

Words in a language follow the same pattern: some of them are much more popular than others. The list of domains that are characterized by such jarring inequalities includes the size of cities (megacities are many times larger than the average city) and number, links, and popularity of web sites (most sites are honored only by the occasional visitor, while others rake in millions). The prevalence of such unequal distributions, where a few elements of a population obtain a disproportionate amount of resources, has been a rich field of inquiry over the last few years. Researchers are looking into the reasons and laws that underlie the emergence of such phenomena.[1]

It is possible that the items in which we are searching exhibit such differences in popularity. Then a search algorithm that will take advantage of the varying popularity of the search items can work much like putting each document that we find at the top of the pile:

1. Search for the item using a sequential search.

2. If the item is found, report that it is found, put it at the front of the list, at its head, and stop.

3. Otherwise, report that the item was not found and stop.

In the following figure, finding E on the list will bring it to the front:

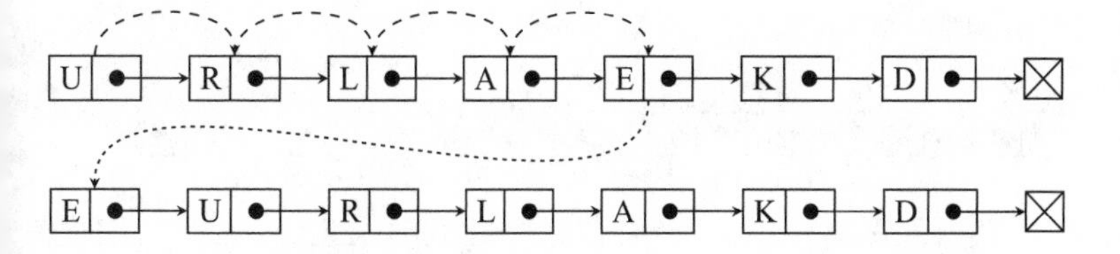

A possible criticism of this *move-to-front* algorithm is that it will promote to the front even an item that we only rarely search for. That is true, but if the item is not popular, it will gradually move toward the end of the list as we search for other items because these items will move to the front. We can take care of the situation, however, by adopting a less extreme strategy. Instead of moving each item we find bang to the front, we can move it just one position forward. This is called the *transposition method*:

1. Search for the item using a sequential search.

2. If the item is found, report that it is found, exchange it with the previous one (if it is not the first one), and stop.

3. Otherwise, report that the item was not found and stop.

In this way, items that are popular will gradually make their way to the front, and less popular items will move to the back, without sudden upheavals.

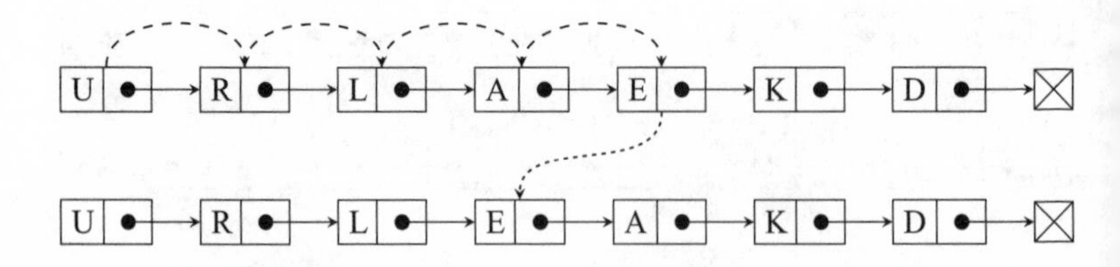

Both the move-to-front and transposition methods are examples of a *self-organizing search*; the name comes because the list of items is organized as we go with our searches and will reflect the popularity of the searched items. Depending on how the popularity ranges among items, the savings can be significant. While with a sequential search we can expect a performance of $O(n)$, a self-organizing search with the move-to-front method can attain a performance of $O(n / lgn)$. If we have about a million items, this is the difference between 1 million and about 50,000. The transposition method can have even better results, but it requires more time to achieve them. That's because both methods require a "warm-up period" in which popular items will show themselves up and make their way to the front. In the move-to-front method, the warm-up is short; in the transposition method, the warm-up takes longer, but then we get better results.[2]

Kepler, Cars, and Secretaries

After the celebrated astronomer Johannes Kepler (1571–1630) lost his wife to cholera in 1611, he set out to remarry. A methodical man, he did not leave things to chance. In a long letter to a Baron Strahlendorf, he describes the process he followed. He planned to interview 11 possible brides before making his decision. He was strongly attracted to the fifth candidate, but was swayed against her by his friends, who objected to her lowly status. They advised him to reconsider the fourth candidate instead. But then he was turned down by her. In the end, after examining all 11 candidates, Kepler did marry the fifth one: 24-year-old Susanna Reuttinger.

This little story is a stretched example of a search; Kepler was searching for an ideal match, among a pool of possible candidates. Yet there was a kink in the process that he was probably not aware of when he started: it might not be possible to go back to a possible match after he had rejected it.

We can recast the problem in more contemporary terms, as looking for the best way to decide which car to buy. We have decided beforehand that we will visit a certain number of car dealerships. Also, our amour propre will not allow us to return to a car dealership after we have walked away from it. If we have declined a car, saving face is paramount, so that we cannot go back and say that we changed our mind. Or perhaps somebody else walked in and bought the car after we left. Be it as it may, we have to make a final decision at each dealership, to buy the car or let go, and not come back.

This is an instance of an *optimal stopping problem*. We have to take an action, while trying to maximize a reward or minimize a cost. In our example, we want to decide to buy the car, when this decision will result in the best car we can buy. If we decide too early, we may settle on a car that is worse than a car we have not seen yet. If we decide too late, we may discover to our chagrin that we saw, but missed, the best car. When is the optimal time to stop and make a decision?

The same issue is usually described in a more callous way as the *secretary problem*. You want to select a secretary from a pool of candidates. You can interview the candidates one by one. You must make a decision to hire or not at the end of each interview, however. If you reject a candidate, you cannot later change your mind and make an offer (the candidate might be too good and thus be snapped by somebody else). How will you pick the candidate?

There is a surprisingly simple answer. You go through the first 37 percent of the candidates, rejecting them all, but keeping a tab on the best one among them as your benchmark. The number 37, which seems magical, occurs because $37\% \approx 1 / e$, where e is Euler's number, approximately equal to 2.7182 (we saw Euler's number in chapter 1). Then you go through the rest of the candidates. You stop at the first of the rest that is better than your benchmark. That will be your pick. In algorithmic form, if you have n candidates:

1. Calculate n / e, to find the 37 percent of the n candidates.

2. Examine and reject the first n / e candidates. You will use the best one among them as a benchmark.

3. Continue with the rest of candidates. Pick the first one that is better than your benchmark, and stop.

The algorithm will not always find the best candidate; after all, the best candidate overall may be the benchmark candidate you identified in the first 37 percent, and that you have rejected. It can be proved that it will find the best candidate in 37 percent (again, $1 / e$) of all cases; moreover, there is no other method that will manage to find the best candidate in more cases. In other words, the algorithm is the best you can do: although it may fail to give you the best candidate in 63 percent of the cases, any other strategy you may decide to follow will fail in more cases than that.

Going back to cars, suppose we decide to visit 10 car dealerships. We should visit the first four and take note of the best offer by these four, without buying. Then we start visiting the remaining six dealerships and we'll buy from the first dealership that gives us an offer better than the one we noted down (we'll then skip the rest). We may discover that all six dealerships make worse offers then the first four that we visited without buying. But no other strategy can give us better odds of getting the best deal.

We have assumed that we want to find the best possible candidate and will settle for nothing less. But what if we can in fact settle for something less? That means that even though ideally we would want the best secretary or car, we can make do with another choice, with which we may be happy, although not as happy had we picked up the best. If we frame the problem like that, then the best way to make our selection is to use the same algorithm as above, but examining and discarding the square root, $\sqrt{n}$, of the candidates. If we do that, the probability that we will make the best choice increases with the number of candidates: as n increases, the probability that we'll pick the best goes to 1 (that is, 100 percent).[3]

Binary Search

We have considered different ways to search, corresponding to different scenarios. A common thread in all these was that the items that we examine are not given to us in any specific order; at best, we order them gradually by popularity in a self-organized search. The situation changes completely if the items are ordered in the first place.

Let's say we have a pile of folders, each one of which is identified by a number. The documents in the pile are ordered according to their identifier, from the lowest to highest number (there is no need for the numbers to be consecutive). If we have such a pile and are looking for a document with a particular identifier, it is foolish to start from the first document and make our way to the last until we find the one we are looking for. A much better strategy is to go straight to the middle of the pile. Then we compare the number identifier on the document in the middle to the number of the document that we are looking for. There are three possible outcomes:

1. If we are lucky, we may have landed exactly on the document that we want. We are done; our search is over.

2. The identifier of the document we are looking for is greater than the identifier of the document we have in our hands. Then we know for sure that we can discard the document at hand *as well as all preceding documents*. As they are ordered, they will all have smaller identifiers. We have undershot our target.

3. The opposite happens: the identifier of the document we are looking for is smaller than the identifier of the document we have in our hands. Then we can safely discard the document at hand *as well as all the documents that come after it*. We have overshot our target.

In either of the last two outcomes, we are now left with a pile that is at most half the original one. If we start with an odd number of documents, say n, splitting n documents in the middle gives us two parts, each with $n / 2$ items (discarding the fractional part in the division):

○ ○ × ○ ○

With an even number of items, splitting them will give us two parts, one with $n / 2 - 1$ items and another one with $n / 2$ items:

○ × ○ ○

We have still not found what we were looking for, but we are much better than before; we have much fewer items to go through now. And so we do. We check the middle document of the *remaining items* and repeat the procedure.

In the figure on the following page, you can see how the process evolves for 16 items, among which we are looking for item 135. We mark out the boundaries inside which we search and the middle item with gray.

In the beginning, the domain of our search is the full set of items. We go to the middle item, which we find out is 384. This is bigger than 135, so we discard it, along with all the items to its right. We take the middle of the remaining items, which turns out to be 72. This is smaller than 135, so we discard it, along with all the items on its left. Our search domain has shrunk to just three items. We take the middle one and find that it is the one we want. It took us only three probes to finish our search, and we did not even need to check 13 of the 16 items.

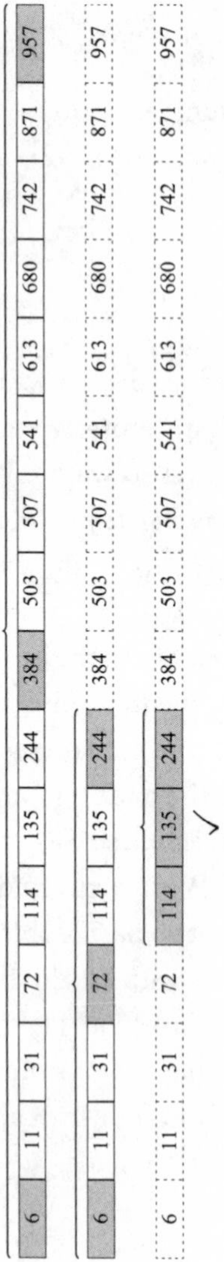

6 11 31 72 114 135 244 384 503 507 541 613 680 742 871 957

6 11 31 72 114 135 244 384 503 507 541 613 680 742 871 957

6 11 31 72 114 135 244 384 503 507 541 613 680 742 871 957

The process will also work if we are looking for something that does not exist. You can see that in the next figure, where we are searching among the same items for one labeled 520.

This time, 520 is greater than 384, so we restrict our search to the right half of the items. There we find that the middle of the upper half is 613, greater than 520. Then we limit our search to just three items, the middle of which is 507. This is smaller than our target of 520. We discard it and now are left with only one item to check, which we discover is not the one we want. So we can finish our search reporting that it was unsuccessful. It took us only four probes.

The method we described is called *binary search* because each time we cut in half the domain of values in which we search. We call the domain of values where we perform our search the *search space*. Using this concept, we can render the binary search as an algorithm comprising these steps:

1. If the search space is empty, we have nowhere to look, so report failure and stop. Otherwise, find the middle element of the search space.

2. If the middle element is less than the search term, limit the search space from the middle element onward and go back to step 1.

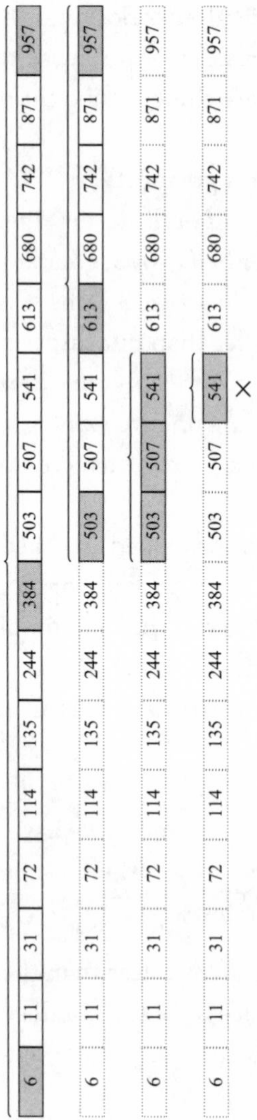

6 11 31 72 114 135 244 384 503 507 541 613 680 742 871 957

6 11 31 72 114 135 244 384 503 507 541 613 680 742 871 957

6 11 31 72 114 135 244 384 503 507 541 613 680 742 871 957

6 11 31 72 114 135 244 384 503 507 541 613 680 742 871 957

×

3. Otherwise, if the middle element is greater than the search term, limit the search space up to the middle element and go back to step 1.

4. Otherwise, the middle element is equal to the search term; report success and stop.

In this way, we divide by two the items that we have to search. This is a divide-and-conquer method. It results in repeated division, which as we have seen in chapter 1 gives us the logarithm. Repeated division by two gives us the logarithm base two. In the worst case, a binary search will keep dividing and dividing our items, until it cannot divide any further, like we saw in the unsuccessful search example. For n items, this cannot happen more than lgn times; it follows that the complexity of a binary search is $O(lgn)$.

The improvement compared to a sequential search, even a self-organized search, is impressive. It will not take more than 20 probes to search among a million items. Viewed from another angle, with a hundred probes we are able to search and find any item among $2^{100} \approx 1.27 \times 10^{30}$, which is more than one *nonillion*.

The efficiency of a binary search is astounding. Its efficiency is probably only matched by its notoriety. It is an intuitive algorithm. But this plain method has proved time and again tricky to get right in a computer program.

For reasons that have nothing to do with the binary search algorithm per se, but rather the way we turn algorithms into real computer code in programming language, programmers have been prey to insidious bugs that have crept into their implementations. And we are not talking about rookies; even world-class programmers have failed to get it right.[4]

To get an idea of where such bugs may lurk, consider how we find the middle element among the items we want to search in the first step of the algorithm. Here is a simple idea: the middle element of the mth and nth elements is $(m + n) / 2$, rounded if the result is not a natural number. This is true, and it follows from elementary mathematics, so it applies everywhere.

Except in computers. Computers have limited resources, memory among them. It is not possible, therefore, to represent all the numbers we want on a computer. Some numbers will simply be too big. If the computer has an upper limit on the size of the numbers that it can handle, then both m and n should be below that limit. Of course, $(m + n) / 2$ is below that limit. But to calculate $(m + n) / 2$, we have to calculate $m + n$ and then divide it by two, *and that sum may be larger than the upper limit*! This is called *overflow*: going beyond the range of allowable values. So you get a bug that you had never thought would bite you. The result will not be the middle value but instead something else entirely.

Do not despair if you find yourself wretched poring over a line of code that does not do what you think it should do. You are not unique. It happens to all; it happens to the best.

Once you know about it, the solution is straightforward. You do not calculate the middle as $(m + n) / 2$ but rather $m + (n - m) / 2$. The result is the same, but no overflow occurs. In retrospect it seems simple. In hindsight, though, everybody is a prophet.

We are interested in algorithms, not programming, here, but let the author share a bit of advice for those who write or want to write computer programs. Do not despair if you find yourself wretched poring over a line of code that does not do what you think it should do. Do not be dismayed if the following day you realize that, indeed, the bug was before your eyes all the time. How could you have failed to see it? You are not unique. It happens to all; it happens to the best.

Binary search requires that the items should be sorted. So to reap its benefits, we should be able to sort items efficiently—which allows us to segue to the next chapter, where we'll see how we can sort things with algorithms.

4

SORTING

The US Constitution postulates that a decennial census should take place in order to apportion taxes and representatives among the several states of the union. The first census following the American Revolution took place in 1790, and a census has been done every ten years since.

In the hundred years since 1790, the United States grew rapidly—from a bit less than 4 million people in the first census, to more than 50 million in 1880. And therein lay a problem: it took eight years to count these people. When the next census year came, in 1890, the population was even bigger. If the count were taken in the same way, it would probably not have been completed before the *following* census of 1900.

At that time, Herman Hollerith, a young graduate from Columbia University's School of Mines (he graduated in 1879, when he was 19), was working for the US

Census Bureau. Aware of the pressing timing problem, he tried to find a way to speed up the census process using machines. Hollerith was inspired by the way conductors used holes punched in railway tickets to record traveler details; he invented a way in which *punched cards* could be used to record census details. These cards could then be processed by *tabulating machines*, electromechanical devices that could read the punched cards and use the data stored in them to make a tally.

Hollerith's tabulating machine was used in the 1890 census and brought down the time required to complete it to six years—when it came out that the US population had grown to approximately 63 million people. Hollerith presented his tabulating machines to the Royal Statistical Society, noting that "it must not be considered that this system is still in an experimental stage. Over 100,000,000 punched cards have been counted several times over on these machines, and this has afforded ample opportunity to test its capabilities."[1] Following the census, Hollerith started a business, called the Hollerith Electric Tabulating System. This company, via a series of renames and amalgamations, evolved into International Business Machines (IBM) in 1924.

Today sorting is so ubiquitous that is largely invisible. Just a few decades ago, offices were full of file cabinets containing labeled folders, and corporate office personnel took care to keep them in the required order, like alphabetic or chronological. By contrast, we can sort the messages in our mailboxes just by clicking, and are able to order them using different criteria such as subject, date, and sender. Our contacts are kept sorted in our digital devices without us taking notice; again, a few years ago we would take pains to make sure we had our contacts organized in our diaries.

Going back to the US census, sorting was one of the first examples of office automation; it is not surprising, then, that it was one of the first applications of digital computers. A lot of different sorting algorithms have been developed. Some of them are not used in practice, but there are still a number of different sorting algorithms that are popular with programmers because they offer different comparative advantages and disadvantages. Sorting is such a fundamental part of what computers do that any book on algorithms will always devote some part to it, yet exactly because there are many different sorting algorithms, their exploration allows us to appreciate an important aspect of the work of computer scientists and programmers. Like toolsmiths, they have a whole toolbox at their disposal. There may be different tools for the same task. Think of different types of screwdrivers. We have slot, Phillips, Allen, and Robertson drivers, to name but a few. Although all of them have the same objective, particular screws require particular drivers. Sometimes we can make do using a slot driver on a cross screw; in general, though, we must use the proper tool for the job. The same with sorting. While all sorting algorithms put things in order, each is more suitable for particular uses.

Before we start exploring these algorithms, let's look at some explanations of what exactly these algorithms do. Sure, they sort stuff, but that really begs the question, What exactly do we mean by *sorting data*?

We assume that we have a group of related data—usually called *records*—that contains some information that is of interest to us. For example, such data could be the emails in our in-box. We want to rearrange these data so that they appear in a specific order that is useful to us. The rearrangement has to take place using some specific feature or features of the data. In our email illustration, we may want to order our messages by delivery date, chronologically, or the sender's name, alphabetically. The order may be ascending, from earlier messages to more recent ones, or descending, from recent messages going back in time. The output of the sorting process must be the same data as the input; in technical terms, this must be a *permutation* of the original data—that is, the original data in different order, but not changed in any other way.

The feature we are using to sort our data is usually called a *key*. A key may be *atomic*, when we consider that we cannot decompose it to parts, or it may be *composite*, when the key consists of more than a single feature. If we want to sort our emails by delivery date, this is an atomic key

Although all of them have the same objective, particular screws require particular drivers. . . . The same with sorting. While all sorting algorithms put things in order, each is more suitable for particular uses.

(we do not care that a date can be broken up in year, month, and day, and may also contain the exact time of delivery). But we may want to sort our emails by the sender's name, *and then* for all the messages from the same sender, order them by delivery date. The combination of date and sender forms the composite key of our sort.

Any kind of feature can be used as a key for sorting, as long as its values can be ordered. Obviously this holds true for numbers. If we want to sort sales data by the number of sales per items sold, the number of sales is an integer. When our keys are textual, such as senders' emails, the ordering that we usually want is lexicographical. Sorting algorithms need to know how to compare our data so as to deduce their order, but any valid way to compare will do.

We'll start our exploration of sorting methods with two algorithms that may be familiar because they are probably the most intuitive and even used by people with no knowledge of algorithms when they have to sort a pile of stuff.

Simple Sorting Methods

Our task is to sort the following items:

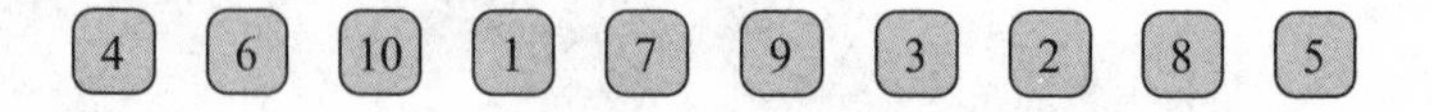

Admittedly, if you take a look at the task, it's pretty trivial; these are the numbers from one to ten. But keeping things simple will allow us to concentrate on the logic of the sorting task.

First, we go through all the items and find the minimum among them. We take it from where we found it and place it first. The minimum of the items is 1, so this must be put into the first position. As this position is already taken, we have to do something with 4, which is currently at the first position; we cannot just throw it away. What we can do is to swap it with the minimum: move the minimum item to the first position and move the item previously in the first position to the position left vacant by moving the minimum. So we go from here, where the minimum is painted black,

to here,

where the minimum is painted white, to indicate that it is in its correct, ordered position.

Now we do exactly the same thing with all the numbers, save for the minimum we found—that is, all the numbers from the second position onward (the gray numbers). We find their minimum, which is 2, and swap it again with the first of the *unsorted* numbers, 6:

Again we do the same. We deal with the items from the third one onward; we find the minimum, which is 3, and swap it with the item currently in the third place, 10:

If we continue this way, item 4 will stay put because it is already in its correct place and we'll go on to place 5 in its sorted position:

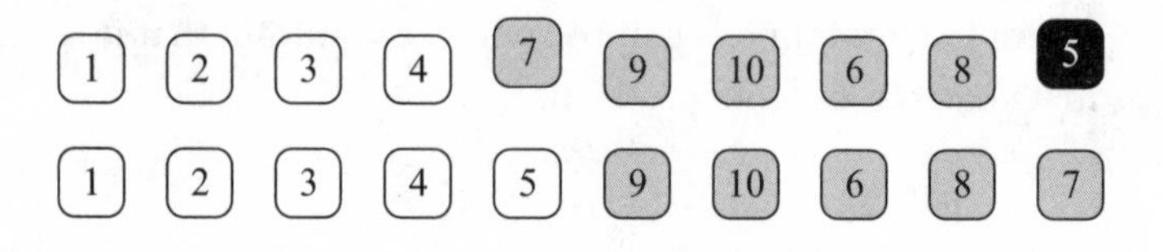

At each point we go through fewer and fewer items to find their minimum. In the end, we'll find the minimum of the last two items, and once we've done that, all our items will be sorted.

This sorting method is called *selection sort* because each time, we select the minimum of the unsorted items and place it where it should be. As all sorting algorithms that we will examine, selection sort has no problem with ties—that is, elements that have the same order. If we find more than one minimum when we examine the unsorted items, we just pick any one of them as our working minimum. We'll find the tied item next time around and put it next to its equal.

Selection sort is a straightforward algorithm. Is it also a good one? If we pay attention to what we are doing, we are going from the beginning to the end of the items that we want to sort, and each time we try to find the minimum of the remaining unsorted items. If we have n items, the complexity of the selection sort is $O(n^2)$. This is not bad in itself; such complexity is not prohibitive, and we can tackle large problems (read: sort a lot of items) in a reasonable amount of time.

The thing is, exactly because sorting is so important, algorithms do exist that are faster than that. So although selection sort is not inherently bad, we usually prefer to use other, more advanced algorithms when we have a lot of items at hand. At the same time, selection sort is not only easy to understand by humans but is also easy to implement on a computer in an efficient way. So it is clearly not of just academic interest; it is really used in practice.

The same can be said for another simple sorting algorithm that we'll describe now. Like selection sort, this is a sorting method that is easy to understand beyond computers. In fact, this is the way we may sort our hand in a card game.

Imagine that you play a game of cards in which you are dealt ten cards (for example, you could be playing Rummy). As you take one card after the other, you want to sort them in your hand. We assume that the card rank, from the lowest to highest, is:

$$2 \quad 3 \quad 4 \quad 5 \quad 6 \quad 7 \quad 8 \quad 9 \quad J \quad Q \quad K \quad A$$

In fact, in many games (and Rummy), the ace can be the lowest- and highest-ranking card, but we'll assume that there is a single order only.

You are dealt each card, so you start with one card in your hand and nine cards to follow:

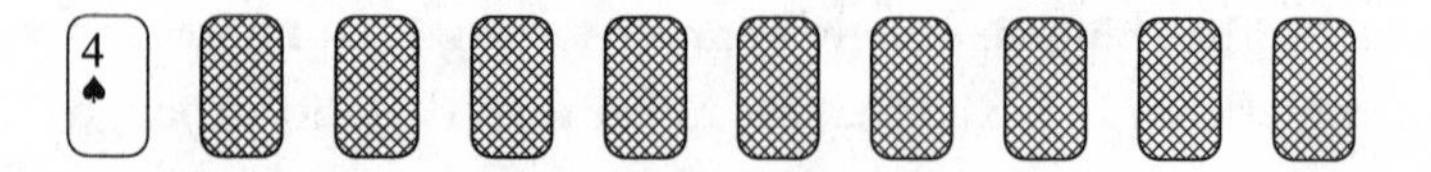

Now you get a second card; it is a six:

Six is fine next to four, so you leave it there and take another card, which turns out to be two:

This time, so as to keep your hand in order, you need to move two to the left of four, thus pushing four and six one position to the right. You do that before you are dealt another card, a three:

You insert the three between the two and four, and see the next card, a nine. This is already in the right place in your hand.

You may continue with your hand—for instance, 7, Q, J, 8, and 5. In the end, you will end up with a sorted hand.

Each new card was inserted in the right place in relation to the previous cards that had been dealt. This way of sorting is called *insertion sort* for that reason, and it works for any kind of objects, not just playing cards.

Like selection sort, insertion sort is straightforward to implement. It turns out that it has the same complexity: $O(n^2)$. It does have a distinct characteristic, though: as in our playing cards example, *you don't need to know the items in advance before you sort them*. In effect, you sort them as you get them. That means that you can use insertion sort when the items to be sorted are somehow streamed to you live. We met this kind of algorithm, which works live as it were, when we discussed the tournament problem in graphs in chapter 2, and we called it an *online algorithm*. If we have to sort an unknown number of items, or if we must be able to stop immediately and provide a sorted list whenever we are suddenly called to do so, then insertion sort is the way to go.[2]

Radix Sort

Let us now return to Hollerith. His tabulating machines did not use selection sort, nor insertion sort. They actually used a precursor of a method still in use today, called *radix sort*. As a tribute to the first machine-enabled sorting application, it is worth spending some time on how radix sort works. It is also interesting because this is a sorting method in which the items to be sorted are not really compared to each other. At least not entirely, as we will see. What's more, radix sort is not just of historical interest, as it performs fantastically. What's not to like in a venerable yet practical algorithm?[3]

The easiest way to see a radix sort is by using playing cards again. Suppose that we have a full deck of cards that has been shuffled and want to sort it. One way to do it is to form 13 piles, one for each rank value. We go through the deck, taking each card and placing it in the respective pile. We'll get 13 piles of four cards each: a pile containing all the aces, another one containing all the twos, and so on.

Then we collect the cards, pile by pile, taking care to put each pile we pick at the bottom of the cards we are collecting. In this way we'll have all the cards in our hands, partially sorted. The first four cards will be aces, the next four twos, and all the way to the kings.

We now create four new piles, one for each suit. We'll go through the cards, taking each card and putting it into the corresponding pile. We'll get four piles of suits. Because the values were already sorted, in each pile we will have all cards of a single suit, in rank order.

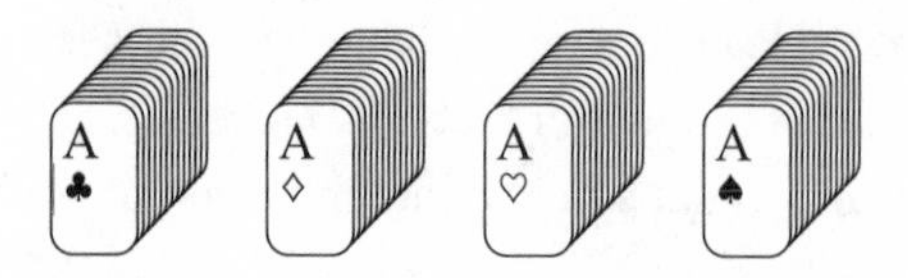

To finish sorting our cards, we only need to collect them pile by pile.

This is the essence of radix sort. We did not sort the cards by fully comparing cards between them. We performed partial comparisons, first by rank, and then by suit.

Of course, if radix sort was applicable only to cards, it would not merit our attention here. We can see how radix sort works with integer numbers. Suppose that we have the following group of integers:

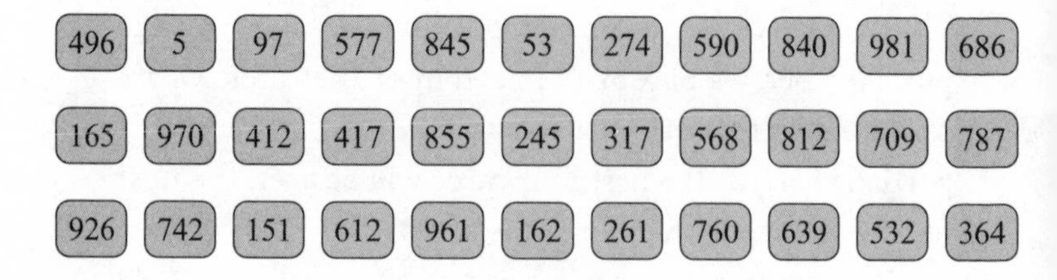

We make sure that all the integers have the same number of digits. So we pad the numbers with zeros on the left if necessary, turning 5 to 005, 97 to 097, and 53 to 053. We go through all our numbers and triage them by their rightmost digit. We use that digit to place them in ten piles:

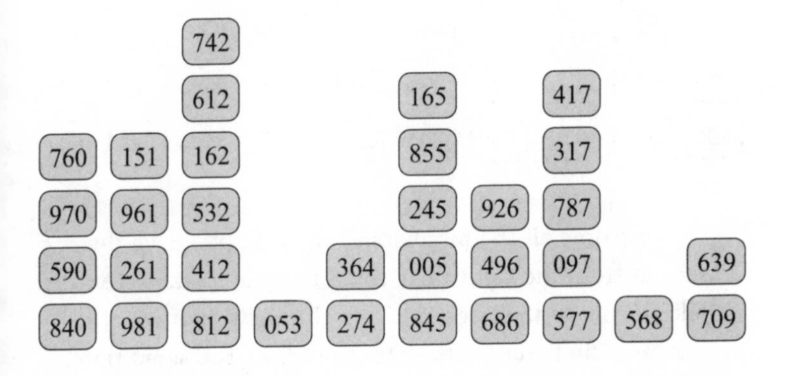

We lightened up the numbers' fill color to indicate that they are partially sorted; each pile contains the numbers with the same rightmost digit. All the numbers in the first pile end in zero, and in the second pile they end in one, up to the last pile, where they end in nine. We now collect the ten piles, starting from the first on the left and adding piles at the bottom (taking care not to shuffle the numbers in any way). Then we redistribute them into ten piles using the second digit from the right and get:

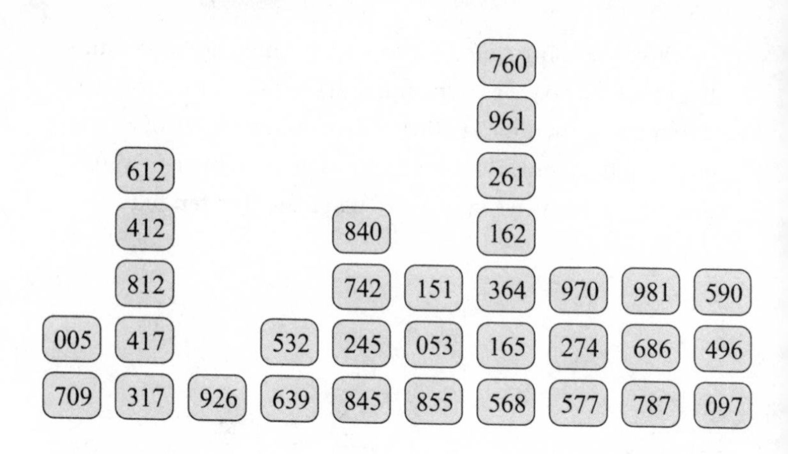

This time all the numbers in the first pile have their second from the right digit equal to zero; in the second pile they have their second from the right digit equal to one, and similarly for the other piles. At the same time, the items in each pile are sorted by their last digit because that's what we did when we piled them the first time.

We finish by collecting the piles and redistributing the numbers, using the third digit from the right this time:

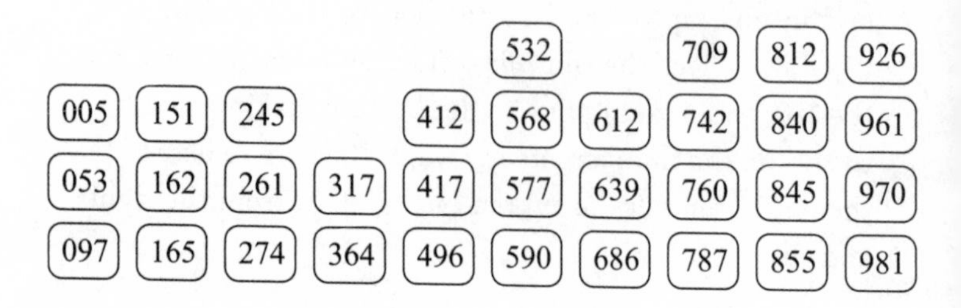

Now the items in each pile start with the same digit and are sorted by their second digit, as a result of the previous piling, and their last digit, as a result of the first piling. To get our sorted numbers, we just collect the piles one final time.

Radix sort can work with words or any sequence of alphanumeric characters as well as integers. In computer science, we call a sequence of alphanumeric characters and symbols a *string*. Radix sort works with strings; the strings may be composed of digits, like in our example, but they may be any kind of strings. The number of piles for alphabetic strings will be equal to the number of distinct characters comprising the alphabet (for instance, 26 piles for English), but the operations will be exactly the same. What is distinctive in radix sort is that even when the strings are composed entirely of digits, we treat them as alphanumeric sequences, not as numbers. If you check how we worked, we did not care for the values of the numbers, but we were working each time with one particular digit from the number, in the same way that we would work by extracting characters from a word, going from the right to left. That is why radix sort is sometimes called a *string sorting method*.

Do not let this fool you and lead you to think that radix sort can order strings while the other sorting methods we present here cannot. All of them can. We can sort strings, as long as the symbols that compose them can themselves be ordered. Human names are strings to computers, and we can sort them because letters are ordered alphabetically and names can be compared lexicographically. The appellation "string sorting" is because radix sort treats all keys, even numbers, as strings. The other sorting methods in this chapter treat numbers as numbers and strings as strings, and work by comparing numbers or strings, as is appropriate. It is only for convenience that we use numbers as keys in our examples in the different sorting algorithms.

The way radix sort works by processing the items to be sorted digit by digit (or character by character) makes it efficient. If we have n items to sort, and the items consist of w digits or characters, then the complexity of the algorithm is $O(wn)$. That is much better than the $O(n^2)$ complexity required by selection and insertion sorts.

And so we come full circle to tabulating machines. A tabulating machine worked in a similar way, sorting punched cards. Imagine that we have a deck of cards where each card has ten columns; punched holes in each column indicate a digit. The machine could recognize the holes in each column, thus figuring out the corresponding digit. An operator put the cards in the machine, and the machine placed the cards in ten output bins depending on their last column—that is, the least significant digit. The operator collected the cards from the output bins, being careful not to mix them in any way, and fed them again into the machine, which this time distributed them into the output bins using their one but last column, the digit next to the least significant one. After repeating the process ten times, the operator could collect an ordered pile of cards. Voilà.

Quicksort

Suppose we have a group of kids milling around in a yard (perhaps at school) and want to put them in line, from the shortest to tallest. Initially we ask them to get in line, which they will do, in whatever order they want:

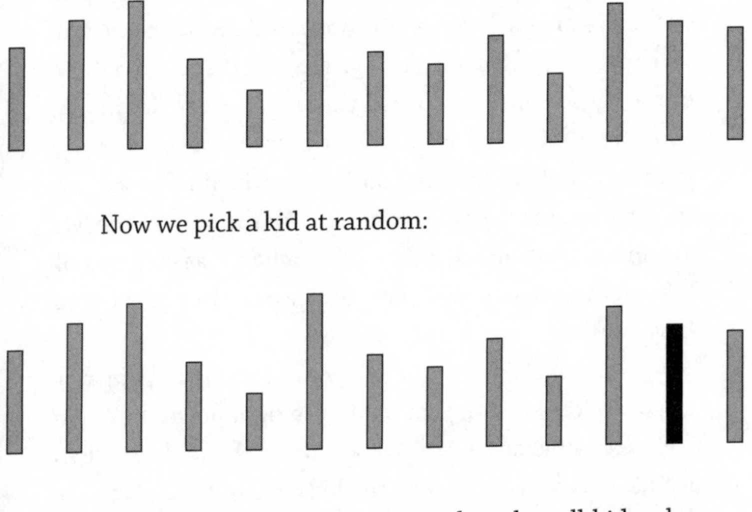

Now we pick a kid at random:

We tell the kids to move around so that all kids who are shorter than the chosen one should move to the left of them and all the rest should move to their right. In the following figure we show where the kid we picked ended up, and you can check that those kids who are taller are to the right and those who are shorter are to the left:

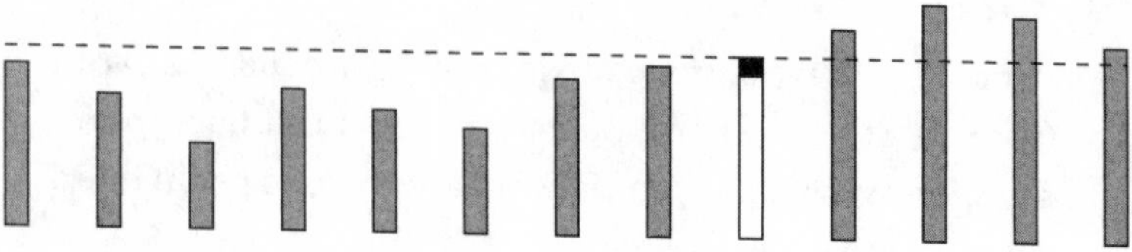

We did not ask the kids to put themselves in the right order. We only asked them to move relatively to the kid we chose. So they formed two groups, on the left and right of the chosen one. The kids in these groups are not in any shorter-to-taller sequence. We do know, however, that *one* kid is certainly in the final position in the line we are trying to form: the very kid we picked. All the kids on the left are shorter and all the kids on the right are at least as tall. We call the kid we picked *pivot* because the rest of the kids have moved around them.

As a visual aid, we will follow the convention of painting white the kids who are put in the right position. When we select a kid as a pivot, we will paint them black; when we have moved the rest of the kids around the pivot, we will use a small black hat to indicate the final position of the pivot (it's white because it's in the right position, with a black top, to indicate that it was the pivot).

Now we shift our attention to one of the two groups, left or right—say the left. Again we pick a pivot in that group at random:

We ask the kids in that group to do the same thing as before: move so that if they are shorter, they move to the left of the pivot and otherwise they should end to the right. We will have again two new, smaller groups, which you can see below. One of them is a group of one, so that kid is in their right place in that trivial group. Then we have the rest of the kids to the right of the second pivot. The second pivot is in the right place, with all the shorter kids to the left, and all the rest to the right. This group to the right extends to the first pivot. We then pick a new, third pivot from that group.

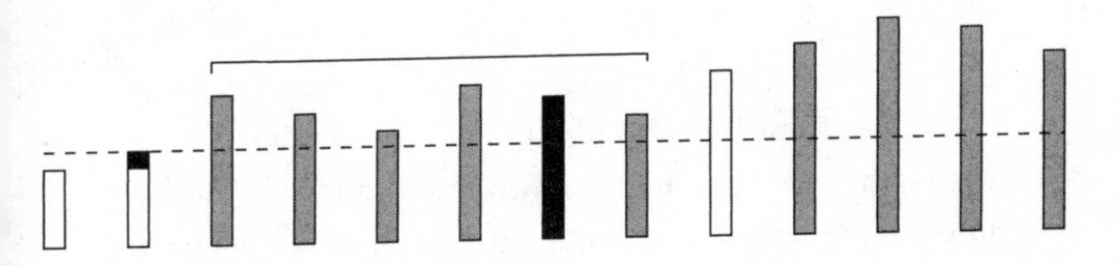

When we tell the kids in the group to move as before, related to how tall they are with respect to the third pivot, two smaller groups will be formed. We focus our attention to the one on the left. We do as before. We pick a pivot, our fourth, and we ask the kids in this group of three to move around it.

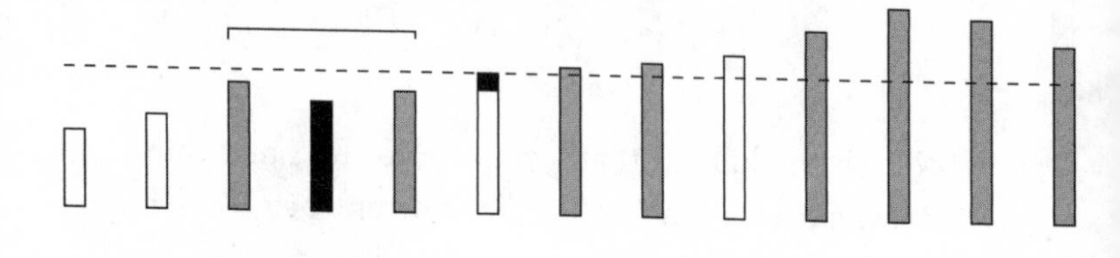

When they do, the pivot ends up being the first of the three, so we have a remaining group of two kids on the pivot's right. We pick one of the pair as a pivot, and the other kid will move, if needed, to their right.

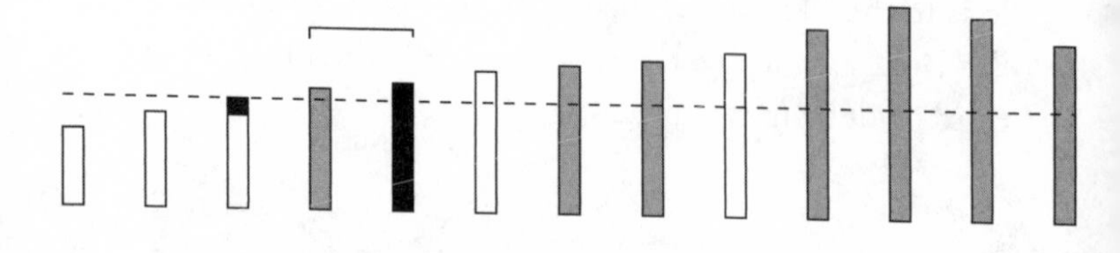

It turns out that this kid does not need to move at all. Right now, we have managed to put about half the kids in order; there are two groups that we had left when we were dealing with previous pivots. We go back to the first of these two groups from the left in order to pick a pivot there and repeat the process.

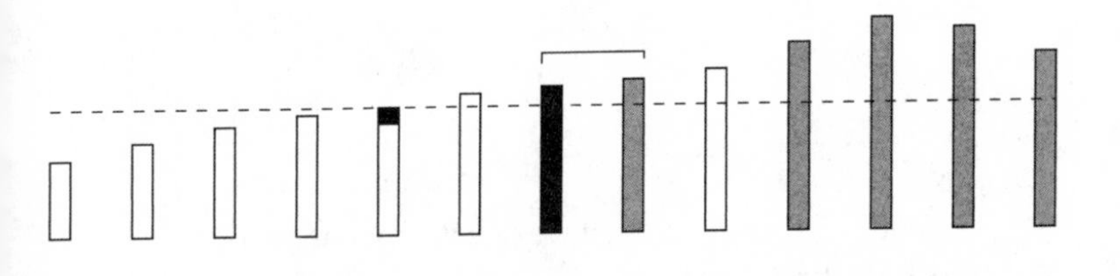

Again, no movement around was necessary and so we go to the last group of unsorted kids to pick a pivot.

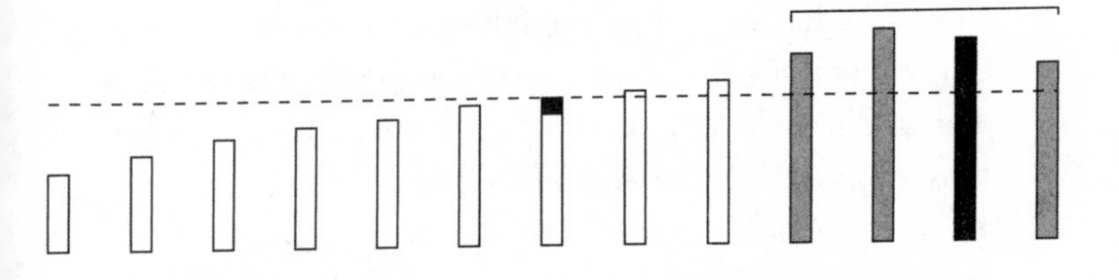

We get a group of one, on the pivot's right, and a group of two, on the pivot's left. We focus on the left group and select one of the two as our last pivot.

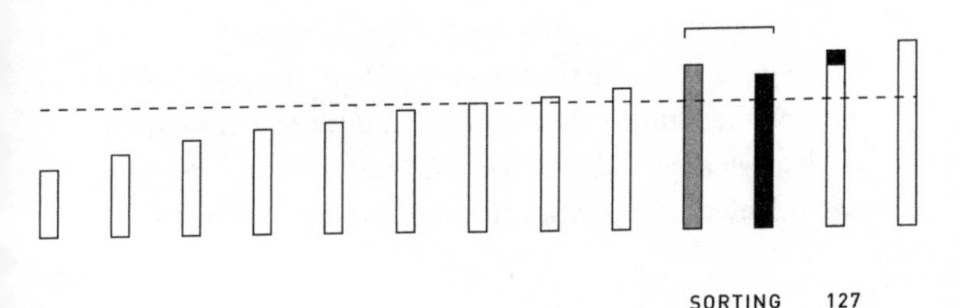

We are done. All the kids are in order of height.

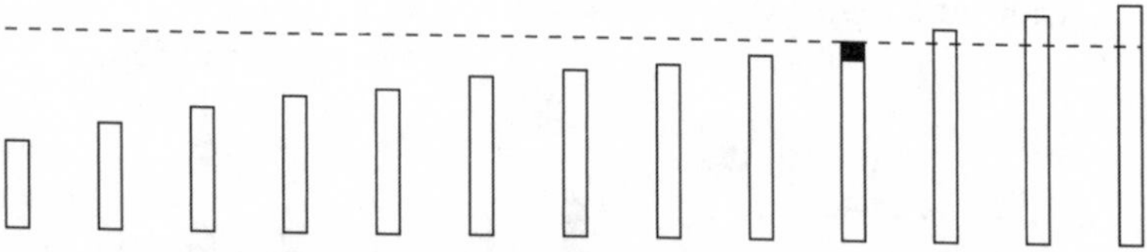

Let's take stock of what we did. We managed to put the kids in order by putting one kid in their right place each time. To do that, we only needed to ask the rest of the kids to move around them. This will always work, of course, not just with kids but also with anything that we may want to sort. If we have a group of numbers that we can sort, we can follow a similar process, picking up a number at random and moving around the rest of the numbers so that those that are smaller end up before our chosen number, and the rest end up after it. We'll repeat the process in the smaller groups that are formed; in the end, we'll have all the numbers in the right order. This is the process that underlies the *quicksort* algorithm.

Quicksort is based on the observation that if we manage to position one element in the correct position with respect to all the rest, whatever that position might be, and then repeat this with the remaining elements, we'll end up with all the elements in their correct positions. If we think back on what we did with selection sort, there we also took each element and positioned it correctly with respect to all the rest, but the element we took was always the minimum of the remaining ones. This is a crucial difference: in quicksort, we should *not* pick the minimum of the remaining elements as our pivot. Let's see what happens if we do so.

If we start again with the same group of kids, we'll get the shortest of all kids as our pivot. That one will go to the beginning of the line, and all the rest will move behind the pivot.

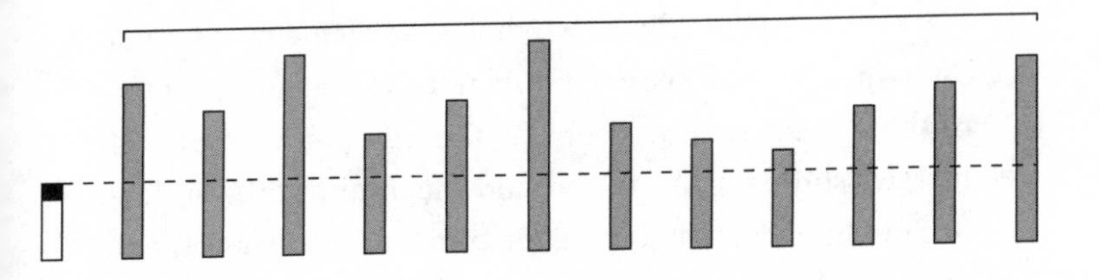

Then we'll get the kid who is immediately taller than the first one and put them second in line. All the rest of the kids will go, again, behind the pivot.

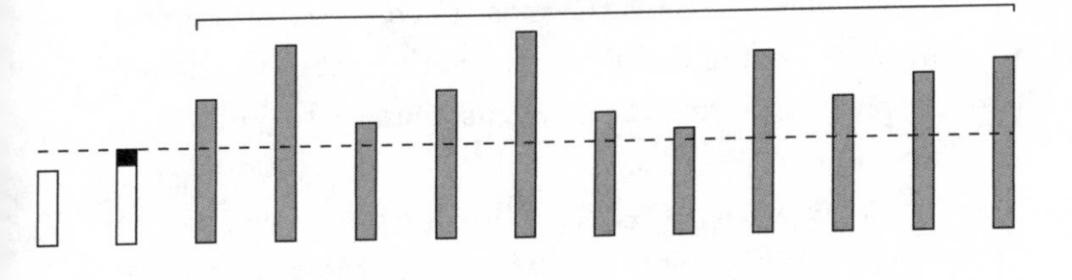

Doing the same thing with the third kid gets us to this point:

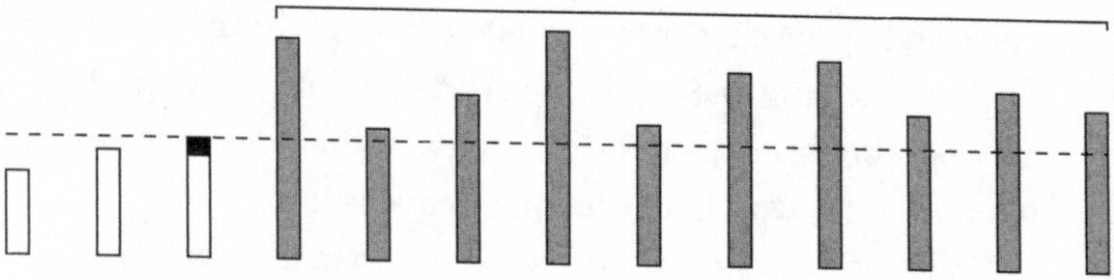

But notice how this looks eerily like a selection sort, as we are filling in the line from the left to the right with the shortest of the remaining kids.

We have not said how we choose an element as a pivot each time. We now see we should not choose the minimum of the elements. First, choosing the minimum requires effort; we should really go and find the minimum each time. Second, it behaves like an algorithm we already know and so there should not be much point in doing it.

The truth is that quicksort is better than selection sort because "normally" (we'll see what normally means shortly) we'll pick as our pivot something that partitions our data in some more equitable way. Choosing the minimum element creates the most unequal partition: nothing on the left of the pivot, and all the rest to the right of the pivot. Each time, then, we just manage to position the pivot itself.

If the partition is better, then we do not just manage to position the pivot. We also manage to position all the elements to the left of the pivot in their correct positions *with respect to the elements to the right of the pivot*. Yes, they are not in their final positions yet. But overall, they are in better positions than before. So we have one element, the pivot, in the best position possible, and the other elements better positioned than before.

This has an important effect on the performance of quicksort: its expected complexity is $O(n \lg n)$, which is way better than $O(n^2)$. If we want to sort 1 million items, $O(n^2)$ works out to 10^{12}, a trillion, while $O(n \lg n)$ is about 20 million.

It all hinges on picking the proper pivot. Searching for a pivot that would partition our data in the best possible way each time does not make sense; it would require searching to find the right pivot, so this would add complexity to the process. A good strategy, then, is to leave it to chance. Just pick a pivot at random and use what you picked to partition the data.

To see why this is a good strategy, let us see why it is not a bad one. It would be a bad one if it led to a behavior like the one we just saw, where quicksort degenerates to selection sort. This would happen if we pick each time as a pivot an item that does not really partition the elements. This can happen if we pick each time the minimum or maximum of the items (the situation is exactly the same). The overall probability of all this happening can be found to be $2^{n-1} / n!$

A probability such as $1 / n!$ is hard to grasp because it is abysmally low. To put it into context, if you take a deck of 52 playing cards and shuffle it randomly, the probability that the deck will end up being in order is 1 / 52! This is about the same as flipping a coin and coming out heads 226 times in a row. When you multiply by 2^{n-1}, things are not improved much. The number $2^{51} / 52!$ is approximately equal to 2.8×10^{-53}. To put the matter in cosmic perspective, the earth is composed of about 10^{50} atoms. If you and a friend of yours were to pick independently an atom from the earth, the probability that you would pick the same atom would be 10^{-50}, actually greater than $2^{51} / 52!$—the probability of pathological quicksort on a deck of cards.[4]

That explains that "normally" we pick a pivot in a more equitable way, as we said above. Excepting a streak of bad luck of cosmic proportions, we do not expect to pick the worst pivot possible each time. The odds actually work better in our favor: it is by picking pivots at random that we expect to get a complexity of $O(n\lg n)$. It is theoretically possible to do worse than that, but the possibility is only of academic interest. Quicksort will be as fast as we expect it to be for all practical purposes.

Quicksort was developed by the British computer scientist Tony Hoare in 1959–1960.[5] It is probably the most popular sorting algorithm today because when implemented correctly, it outruns all others. It is also the first algorithm that we see whose behavior is not entirely deterministic. Although it will always sort correctly, we cannot guarantee that it will always have the same run-time performance. We can guarantee that it is extremely unlikely that it will exhibit pathological behavior. This is an important concept, which brings us to the so-called *randomized algorithms*: those algorithms that use an element of chance in their operation. This runs contrary to our intuition; we expect algorithms to be the ultimate deterministic beasts, slavishly following the instructions we lay down for them on a preordained path. And yet randomized algorithms have blossomed in recent years, as it has turned out that chance can help us solve problems that remain intractable to more standard approaches.[6]

Merge Sort

We've met radix sort, which essentially sorts items by distribution: in each round through the data, it places each item in a correct pile. Now we'll meet another sorting method, which sorts item by *merging* stuff together instead of splitting them apart. The method is called *merge sort*.

Merge sort starts by admitting to a limited capability for sorting; imagine that we are unable to sort our items if they are given to us in any random arrangement. We are only able to do the following: if we are given two groups

Randomized algorithms have blossomed in recent years, as it has turned out that chance can help us solve problems that remain intractable to more standard approaches.

of items, and each group is already sorted, we can merge them together and get a single, sorted group.

For example, say we have the following two groups, one per row (although in our example the two groups have the same number of items, there is no need for the groups to be equal in size):

As you can see, each of the two groups is already sorted. We want to merge them in order to create a single sorted group. This is really simple. We check the first item of both groups. We see that 15 is smaller than 21, so this will be the first item of our third group:

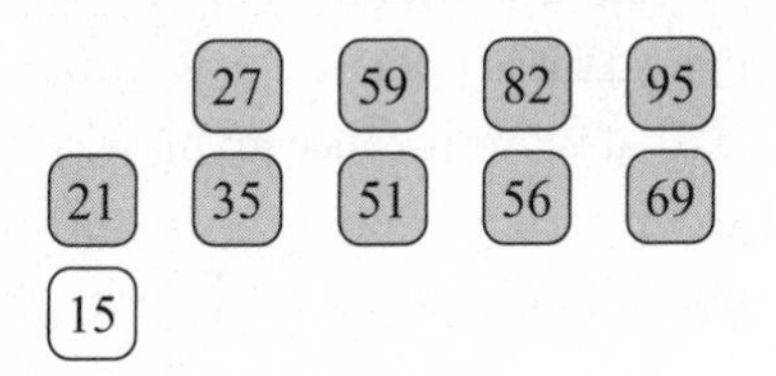

We examine again the first elements of the two groups, and this time 21 from the second group is smaller than 27 from the first group. So we take it and append it to the third group.

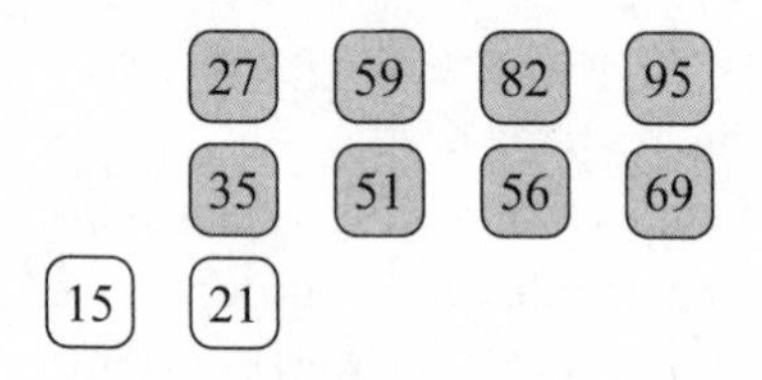

If we continue in this way, we'll take 27 from the first group and then 35 from the second group, adding them to the end of the third group:

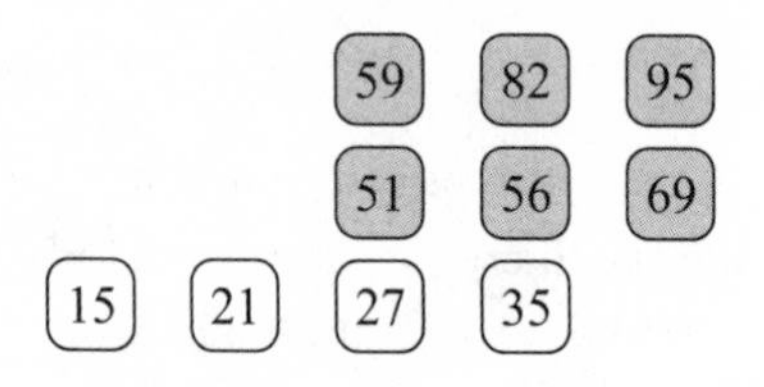

Now 51 is smaller than 59, and 56 is smaller than 59. As we already have moved 35 from the second group to the third, in the end we'll have moved three items in a row from the second group to the third. That is fine because in this way we keep items in the third group sorted. There is no reason why the two first groups should diminish at the same rate.

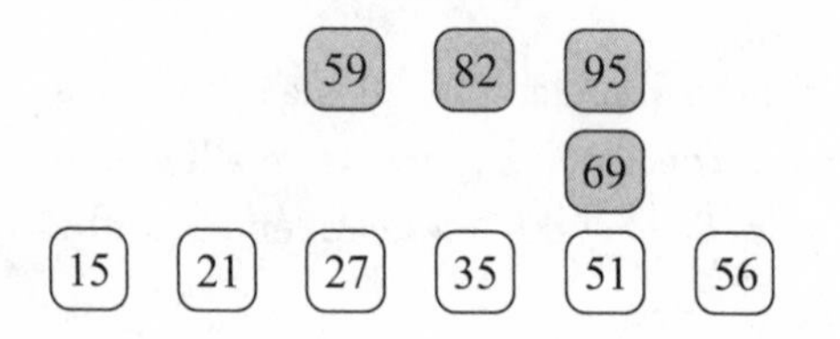

We return to the first group, as 59 is smaller than 69, so we add it to the third group:

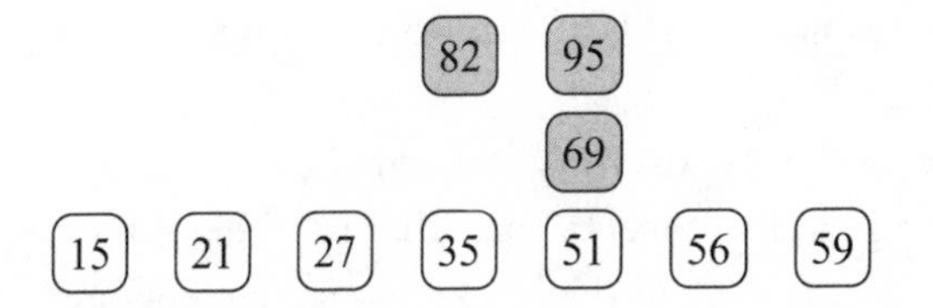

Next, by moving 69 to the third group we empty the second group completely:

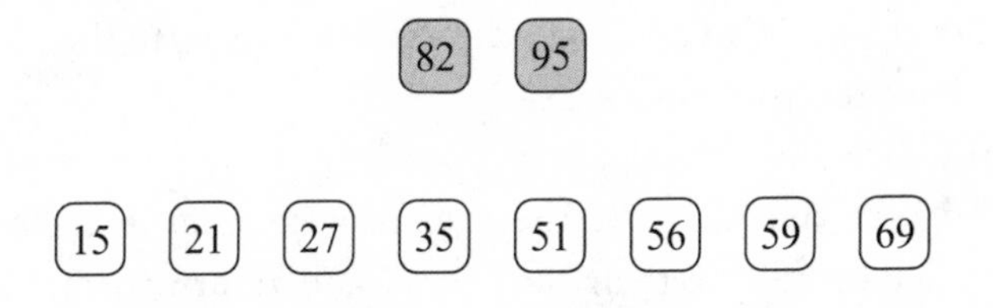

We finish by moving the last remaining elements of the first group to the third group—they are definitely larger than the last element of the third group or otherwise we would not have moved it there previously. Our items are completely sorted now:

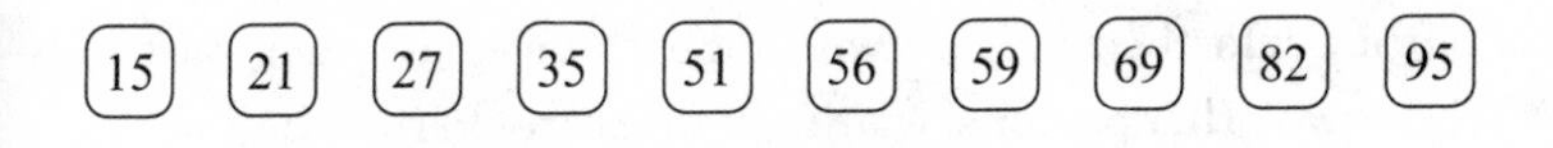

It's nice to have a way of producing a sorted group from two sorted groups, but this does not seem to solve our problem of sorting a single group of unsorted items. It is true it does not, yet it is an important component of the solution.

Imagine now that we have a group of people. We give to one of them a group of items to sort. That person does not know how to sort them, but they do know that if somehow they had two sorted parts of the items, they could produce a final sorted group. So what they do is this: they split the group in two and pass it on to two other people. They say to the first of them, "Take this group and sort it. Once you are done, return it to me." They say the same thing to the second person. Then they wait.

Although our first point of contact does not know how to sort the items, if the two new contacts manage somehow to sort their own parts and return them, then the first person would return to us the final, completely sorted group. But the two other contacts know no more than our initial contact—they don't know how to sort but rather only how to merge sorted stuff using the algorithm above—so has anything really been achieved?

The answer is yes, provided that they do the same: they split their part in two, and each delegates their part to two other persons, waiting for *them* to do their bidding and provide them with two sorted parts.

This seems like the ultimate pass-the-buck game, but look at what happens if we try to see it unfold with an example. We start with the numbers 95, 59, 15, 27, 82, 56, 35, 51, 21, and 79. We give them to Alice (A), who splits them in two, and passes them to Bob (B) and Carol (C). You can see that in the first level of the upside-down tree below:

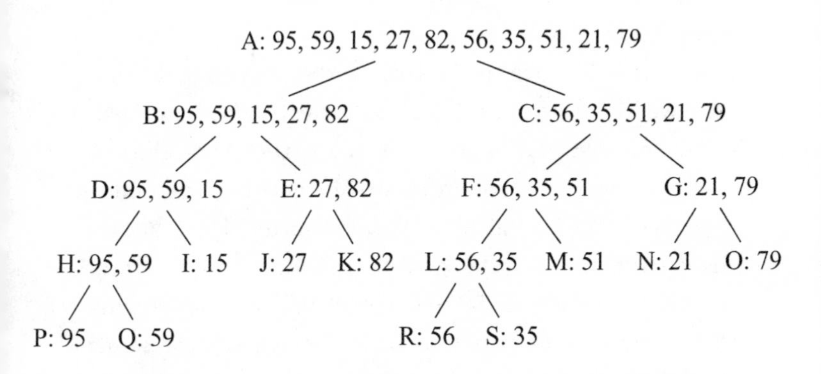

Then Bob splits his numbers into two, and passes them on to Dave (D) and Eve (E). Similarly, Carol splits her numbers, and passes them on to Frank (F) and Grace (G). Our cast of characters continue passing the buck. Dave divides his numbers to Heidi (H) and Ivan (I); Eve distributes her two numbers to Judy (J) and Karen (K); Frank and Grace split to Leo (L) and Mallory (M) and Nick (N) and Olivia (O), respectively. Finally, Heidi splits her pair to Peggy (P)

and Quentin (Q), while Leo splits his pair to Robert (R) and Sybil (S).

The people at the leaves of the tree have really nothing to do. Peggy and Quentin receive a number each, and they are told to sort it. But a single number is sorted by definition: it is in order with itself. So Peggy and Quentin just give their number back to Heidi. Also, Ivan, Judy, Karen, Robert, Sybil, Mallory, Nick, and Olivia return the numbers they received.

Now let's move to the tree on the next page. In this tree we'll move from the leaves, at the top (so this looks like a normal tree, not upside down), to the root at the bottom. Let's concentrate on Heidi. She gets back two numbers, each one of which is (trivially) sorted. Heidi knows how to merge two sorted groups to produce a single group so she can use 95 and 59 to make 59, 95. She then returns this sorted group of two to Dave. Leo will act the same: he will get 35 and 56, which are already sorted (by themselves), and knows how to put these two in order and create 35, 56, which he returns to Frank.

Dave, who was clueless about the numbers 95, 59, 15 that he had initially received, now gets 59, 95 from Heidi and 15 from Ivan. Both of these groups are already sorted, which means that Dave can merge them to create 15, 59, 95. In the same way, Frank gets 35, 56 from Leo and 51 from Malory, and can produce 35, 51, 56.

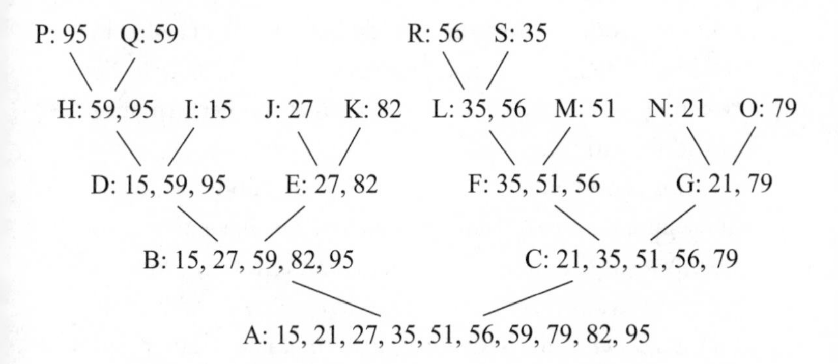

If everybody acts in the same way, when the numbers reach Alice, she will get two sorted lists, one from Carol and one from Bob. She will merge them to create the final sorted list.

These two trees are the essence behind merge sort. We delegate the sorting as much as we can, to the point that no sorting can take place because lone items are already sorted by definition. Then we merge larger and larger groups, until we absorb all elements in a single, final, sorted group.

The smarts that we require from our characters is minimal. You can see in the first tree that Eve got from Bob a group of numbers that as it happened was already sorted: 27, 82. It does not matter. She does not stop to check whether they need sorting or not—and we don't want her to because such a check would take time. She just splits and passes them down. She will get them back and merge them to produce what she already got. That's all right; in the large scheme of things, this gratuitous pas de trois between Eve, Judy, and Karen won't affect the performance of the algorithm.

The complexity of merge sort is as good as that of quicksort, $O(nlgn)$. That means that we have two algorithms with the same performance. In practice, programmers may choose one or the other depending on additional factors. Usually quicksort programs run faster than merge sort ones because their concrete implementation in a programming language is faster. Merge sort splits the data before merging them, which means that they can be parallelized, so that vast amounts of data can be sorted by a computer cluster, where each computer acts like our human sorters above.

Merge sort is as old as computers. Its inventor was a Hungarian American, Neumann János Lajos, better known under his American name, John von Neumann (1903–1957). In 1945, he wrote a manuscript, in ink, 23 pages long, for one of the first digital computers, the Electronic Discrete Variable Automatic Computer, or EDVAC for short. At the top of the first page, the phrase "TOP SECRET" was penciled in (and later erased), as work on computers was classified in 1945 due to its connections with the military. The subject of the paper was a nonnumerical application of computers: sorting. The method that von Neumann described was what we now call merge sort.[7]

5

PAGERANK

If you are below a certain age, the words HotBot, Lycos, Excite, AltaVista, and Infoseek mean nothing to you, or if they do mean something, they probably do not mean search engines. Yet all of them were vying for our attention at some point or other, trying to get us to use them as the gateway to the web.

This is history now, as the search engine landscape is dominated by two services, Google, run by Alphabet, and Bing, run by Microsoft. The explosion of many competing solutions in a new market, and their subsequent consolidation, is a pattern that we have witnessed in many industries in history. What is remarkable in the search engine space is that we know that a large factor in the evolution is the phenomenal success of Google, which in turn was based on an algorithm that its founders invented. The founders were Larry Page and Sergey Brin, doctoral candidates at Stanford University, and they named their algorithm PageRank, after Page (and not after "page" and rank, as one might expect).

Before we embark on a description of PageRank, we need to understand what exactly search engines do. This is actually two things. First, they crawl the web, reading and indexing all the web pages they can come across. In this way, when we type in a search term, search engines look into the data they have stored on the crawled web pages and find the ones that match our query. So if we search for "climate change," the search engines will search through the data they have amassed to find the web pages that contain this search term.

If our search term describes a popular topic, the results can be numerous. At the time of this writing, the query "climate change" on Google returns more than 700 million results; this number may be different when you read these lines, but you get an idea of the scale. This brings us to the second thing that search engines do. They must present the search results so that those that are more pertinent to what we are looking for appear first, and those that are less likely to interest us appear later. If you are trying to learn the facts about climate change, you would expect to see results from the United Nations, National Aeronautics and Space Administration (NASA), or Wikipedia come up on top. You would be rather surprised if the top result was a web page explaining the view of the Flat Earth Society on the topic. From the hundreds of millions of web pages that may be related to your query, many will be trivial; others may be bloviating, and yet others will be utter nonsense. You want to hone in on those that are to the point and authoritative.

When the Google search engine arrived on the scene (the author is old enough to remember), people (the author included) started switching to the newcomer from other, older, now-extinct search engines because its results were better and they arrived faster. It also helped that the Google web page was plain, containing only relevant information, instead of being flush with all sorts of paraphernalia, which had been the fashion. We'll leave aside the second factor, illuminating though it is (Google understood that users cared for good and fast search results, not for bells and whistles), and deal with the first. How could Google deliver better results than the others, fast?

If the web were small, we could create a catalog of it, and have editors to curate the catalog and assign an importance to its entries—the web pages. But the scale of the web precludes such an approach, although there were such attempts before it became obvious that the size of the web would make this an impossible task.

The web consists of web pages, linked to each other through links. We call these links *hyperlinks*; text that contains such cross-references to other parts of the text

If you are trying to learn the facts about climate change, . . . you would be rather surprised if the top result was a web page explaining the view of the Flat Earth Society on the topic.

or other texts is called *hypertext*. The notion of hypertext predates the web. The first description of a system of organizing knowledge by interlinking documents was written by the US engineer Vannevar Bush and appeared in 1945 in the *Atlantic*. The World Wide Web, or simply the web as it became known, was developed by the British computer scientist Tim Berners-Lee in the 1980s. Berners-Lee was working at CERN, the European Organization for Nuclear Research, outside Geneva, Switzerland, and wanted to create a system to help scientists share documents and information. They could do that by making their documents available online and also adding links from their documents to other documents that were available online. The web has grown, and continues to grow, organically by people adding new pages. Authors of web pages write the content of the pages and link to existing pages that are relevant to the content of the pages they write.

Imagine you are the author of an online article that provides an overview of the effects of climate change in your country. In the article, as you introduce the topic, you may want to let your readers navigate to a web page that you believe is an authoritative source on the matter, so you add a link to that web page. In this way you help your readers by allowing them to delve deeper into the subject, while at the same time you add gravitas to your own writing because you substantiate your statements by those of another web page that you trust.

There are many people like you, writing their own online articles on the effects of climate change in their countries or regions. Each one of them may also want to link to what they believe is an authoritative source on the topic. Hyperlinks will emanate from these online articles to point to relevant sources of information.

The reason why NASA might come up on top in a search for climate change is that lots of authors, each one writing their own article, decided to place a hyperlink to the NASA web page on climate change. Authors made their own choices individually, but it is likely that many chose the same page, such as, for instance, NASA's page. It therefore makes sense that this page on climate change should be judged important, relative to other web pages.

The whole system acts as a kind of democracy. Authors of web pages link their pages to other pages. The more links that a web page accrues, the more authors judged it important enough to link to it from their own page, and thus the more important it becomes overall.

There is, though, a conceptual difference from democracy as we usually practice it. Not all of these articles that are written are equal. Some of them appear on more prestigious web sites than others. An article on a blog that is read by a handful of people carries less weight than an article in an online publication that rakes in hundreds of thousands of readers. This indicates that we should not consider just the number of links pointing to a web page

The whole system acts as a kind of democracy. Authors of web pages link their pages to other pages. The more links that a web page accrues, . . . the more important it becomes overall.

in order to gauge its importance. Who is pointing to a web page is also significant, not just how many. It is reasonable to expect that a link from a prestigious web page carries more weight than a link from an obscure site. Although you should not judge a book by its cover, an endorsement by a prominent author is more important than a good review by an unknown reviewer. Every link from one page to another page acts as an endorsement from the first page to the second, and the weight of the endorsement depends on the status of the endorser. At the same time, if a page links to many other pages, its endorsement should be divided, as it were, among the pages that receive it.

The set of pages linked by hyperlinks forms an enormous graph, containing billions of pages and many more links between them. Every web page is a node in the graph. Every link from one page to another is a directed edge in this huge graph. The fundamental insight behind PageRank is that following the reasoning we have just outlined, we can use the structure of the web graph to give us the importance of each web page. To be more precise, we can get the importance of each page through a number. This number, which we will call its pagerank, will measure the significance of a web page related to the other web pages. The more important a web page is, the higher its pagerank will be. The PageRank algorithm follows the ramification of this insight on a humongous scale, on the graph representing the whole web.

The Basic Principles

When we are on a web page, the links on that page point to other pages that are relevant to the page we are currently browsing. The very existence of the link indicates that the web page at the end of the link is important—otherwise the author of the web page would not link to it in the first place. Consider the example graph below, representing a small set of web pages that link to each other:

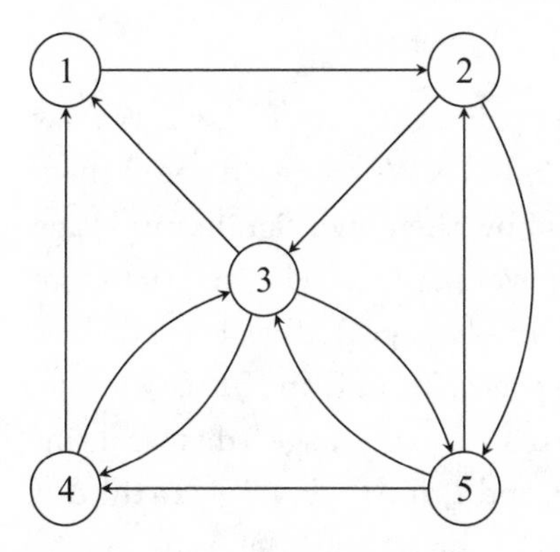

In such a graph, we call the links that point to a web page *backlinks*; by extension, we will also call the pages that point to a web page *backlinks*. In this way, the backlinks of web page 3 are the edges pointing to it, its incoming edges, as well as the nodes from which they emanate: web pages 2, 4, and 5. As in this chapter we will be concerned with graphs that are made up of web pages, we will be using the terms “node” and “page” interchangeably.

We will build an algorithm for finding the importance of each web page based on two basic principles:

1. The importance of a web page depends on the significance of the web pages that link to it—that is, on the importance of its backlinks.

2. A web page divides its importance evenly over the web pages to which it links.

Say we want to find the importance of page 3. We saw that its backlinks are 2, 4, and 5. We take each one of them in turn and assume we know their own significance. Page 2 divides its importance over pages 3 and 5, and therefore will give half its importance to page 3. Page 4 also divides its importance over two pages, 3 and 1, and hence will give half its significance to page 3. Finally, page 5 divides its importance over pages 2, 3, and 4, and thus will give a third of its importance to page 3. To save typing, let us denote by $r(P_i)$ the importance of page i; r will stand for rank. Then the importance of page 3 will be:

$$r(P_3) = \frac{r(P_2)}{2} + \frac{r(P_4)}{2} + \frac{r(P_5)}{3}$$

In general, if we want to find out the importance of a certain web page and we know the importance of each backlink, it is easy to find what we are looking for: divide the importance of each backlink page by the number of web pages it links to and add the result to the contributions of the other backlinks of the page.

You may think of the calculation of importance as a voting contest between web pages. Each voting page has some significance, which it can use as a vote for those web pages that it deems important. If it considers only one web page as important, it just gives its vote to that web page. But if it considers more than one web page as significant, then it splits its vote and gives a part of the vote to each of these web pages. Therefore, if a web page wants to vote three web pages as being important, it will give to each one of them one-third of its vote. To which pages will a web page apportion its vote? To those at the end of its hyperlinks—that is, to those to which it links. And how is the importance of a web page derived? From the importance of its backlinks.

The two principles do endow some aura of democracy to the ranking of web pages. There is no single authority that decides what is most significant. A web page is important if other web pages think it is important, and they vote with their links. In contrast with the one person, one vote principle that holds in most real-world elections, however, not all web pages have equal votes here. The votes of a web page depend on how important it is—which, again, is determined by the other web pages.

This may seem like casuistry because in effect it tells us that to find the importance of a web page, we must find the importance of its backlinks. If we follow the same reasoning, to find the importance of each of its backlinks, we must find the importance of that backlink's backlinks. Then the process seems to regress more and more, from backlinks to backlinks, and in the end, we are left without knowing how to calculate the significance of the web page from where we started. Worse, we may find out that we run in circles. In our example, to calculate the importance of page 3, we need the importance of each of pages 2, 4, and 5. To calculate the importance of page 2, we need the importance of page 1 (and page 5, but let us leave that aside for a bit). To calculate the importance of page 1, we need the importance of page 4, and to find that, we need to know the importance of page 3. We are back where we started.

An Example

To see how we get out of the problem, let us assume that before we begin calculating the importance of the web pages, we give them all equal significance. In terms of our voting metaphor, we give each web page exactly one vote. When the voting starts, each one of the pages will vote in the way we described, spreading its vote to the pages to which it links. Each page will then receive votes from all its backlinks. The transfer of votes will look like this:

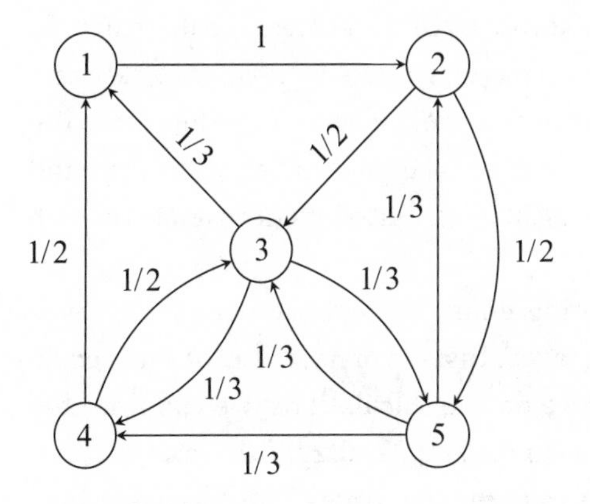

Page 1 sends its vote to page 2, the only page it links to. Page 2 divides its vote into two parts, and sends 1 / 2 to page 3 and 1 / 2 to page 5. Page 3 divides its vote into three parts and sends 1 / 3 to each of pages 1, 4, and 5. Pages 4 and 5 vote using the same method.

Once voting is over, each page will calculate the total from the sum of the votes, or fractions of the votes, it has received from its backlinks. For example, page 1, having received votes from pages 3 and 4, will have 1 / 2 + 1 / 3

$= 5 / 6$ votes, while page 3, having received votes from pages 2, 4, and 5, will have $1/2 + 1/2 + 1/3 = 4/3$ votes. We see that page 1 decreased its share of votes compared to where it started, while page 3 increased it.

Now let us change the setup a little bit. Instead of giving each page one vote before the voting starts, we give each page 1 / 5 of a vote so that all votes sum up to one. In general, if we have n pages, we give $1 / n$ votes to each one of them. The rest of the process is exactly the same. The overall importance of all web pages is equal to one, and the importance is again distributed evenly over all the web pages.

After the voting ends, the importance of each web page will have changed. Instead of having all of them equal to $1/5 = 0.2$, if we do the calculations, we will find that they will be equal to 0.17, 0.27, 0.27, 0.13, and 0.17 for each of the pages in turn. Web pages 2 and 3 have gained in importance, while web pages 1, 4, and 5 have lost importance. The total significance of all web pages sums up to one.

We can now start another voting round, with exactly the same rules. The pages will spread the votes they have gathered to the pages to which they link. At the end of this second round, each page will count its votes to determine its standing in terms of accumulated importance. After the calculations, the new importance values will be 0.16, 0.22, 0.26, 0.14, and 0.22.

We'll do exactly the same process again. In fact, we'll repeat the voting again and again. If we do that, the votes—that is, the importance apportioned to each page—will evolve as in the following table, which shows the initial values and results after each voting round:

Round	Page 1	Page 2	Page 3	Page 4	Page 5
start	0.20	0.20	0.20	0.20	0.20
1	0.17	0.27	0.27	0.13	0.17
2	0.16	0.22	0.26	0.14	0.22
3	0.16	0.23	0.26	0.16	0.20
4	0.17	0.22	0.26	0.15	0.20
5	0.16	0.23	0.25	0.15	0.20
6	0.16	0.23	0.26	0.15	0.20

If we go on to perform another, seventh voting round, we'll discover that the situation will remain unchanged with respect to the sixth voting round. The votes, and therefore the importance of the web pages, will remain the same. This then gives us our final result. The ranking of the web pages is that page 3 is the most important, followed by page 2, then page 5, then page 1, and last comes page 4.

Let's step back and reflect on what we did. We started with two principles that give us rules for calculating the importance of a web page, provided we know the importance of each of its backlinks. Before we start, we set up all

n web pages with equal importance, equal to $1 / n$. Then we calculate the significance of each web page by summing the shares it gets from its backlinks. This gives us new values for the significance of each web page, different from the $1 / n$ value from where we started. We repeat the process beginning with these values. We find another set of values. After a number of repetitions of this process, we found that the situation stabilized: the measure of importance would not change from one repetition to the next. At this point we called it a stop and reported the values that we found.

The question of course is whether the approach that we have just described works in general and not in the particular example that we chose. Moreover, does it produce sensible results?

The Hyperlink Matrix and Power Method

The method of calculating the importance of a page from the importance of its backlinks has an elegant formulation. We start from the graph that describes the links between our web pages. We can represent a graph by using a *matrix* of numbers, which we call its *adjacency matrix*. The construction is straightforward. We create a matrix with as many rows and columns as the nodes in the graph. Then we put one for each intersection that corresponds to a link and zero for all other intersections. The adjacency matrix for our example is:

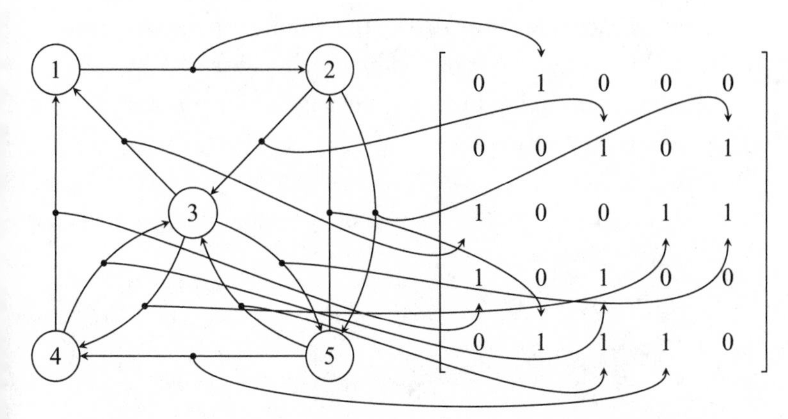

We can also represent the importance of the web pages using a single row or *vector*:

$$[r(P_1) \quad r(P_2) \quad r(P_3) \quad r(P_4) \quad r(P_5)]$$

As we now get into the nuts and bolts of the PageRank algorithm, we'll start using the term pagerank to refer to the significance of a web page. You will see that the term will be justified as we will be able to derive a ranking, in terms of importance, of all the pages on the web. As our row contains all the pageranks, we will call it the *pagerank vector* of our graph.

The importance of each web page is divided over the pages to which it links. Now that we have the adjacency matrix at hand, we can do that by going to each row and dividing each one in the row by the number of ones in that row. This is equivalent to dividing each page's vote by the number of outgoing links to that page. If we do that, we get the following matrix:

$$\begin{bmatrix} 0 & 1 & 0 & 0 & 0 \\ 0 & 0 & 1/2 & 0 & 1/2 \\ 1/3 & 0 & 0 & 1/3 & 1/3 \\ 1/2 & 0 & 1/2 & 0 & 0 \\ 0 & 1/3 & 1/3 & 1/3 & 0 \end{bmatrix}$$

We call this matrix the *hyperlink matrix*.

If we look carefully at the hyperlink matrix, each column shows how the importance of a page is derived from the pages that link to it. Take the first column, which relates to the importance of page 1. This page takes its significance from pages 3 and 4. Page 3 gives 1 / 3 of its importance to page 1 because it links to three pages, and page 4 gives 1 / 2 of its importance to page 1 because it links to two pages. Page 1 receives zero significance from the other pages in the graph because they do not link to it. We can express this as:

$$r(P_1) \times 0 + r(P_2) \times 0 + \frac{r(P_3)}{3} + \frac{r(P_4)}{2} + r(P_5) \times 0 = \frac{r(P_3)}{3} + \frac{r(P_4)}{2}$$

But this is exactly the definition of $r(P_1)$, the pagerank of page 1. We got the pagerank by summing the products of the elements of the pagerank vector with the corresponding elements of the first column of the hyperlink matrix.

Let's see what is happening if we take the pagerank vector and sum the products of its elements with the corresponding elements of the second column of the hyperlink matrix:

$$r(P_1)\times 1 + r(P_2)\times 0 + r(P_3)\times 0 + r(P_4)\times 0 + \frac{r(P_5)}{3} = r(P_1) + \frac{r(P_5)}{3}$$

That is exactly the definition of $r(P_2)$, the pagerank of page 2. The sum of the products of the elements of the pagerank vector with the contents of the third column of the hyperlink matrix will similarly give us $r(P_3)$, the pagerank of page 3:

$$r(P_1)\times 0 + \frac{r(P_2)}{2} + r(P_3)\times 0 + \frac{r(P_4)}{2} + \frac{r(P_5)}{3} = \frac{r(P_2)}{2} + \frac{r(P_4)}{2} + \frac{r(P_5)}{3}$$

You can verify that using the fourth and fifth columns of the hyperlink matrix we'll get $r(P_4)$ and $r(P_5)$, respectively. This operation—of summing the products of the elements of the pagerank vector with the contents of each column of the hyperlink matrix—is actually the product of the pagerank vector with the hyperlink matrix.

Unless you are familiar with matrix operations, this may be confusing because we usually talk about the product of two numbers, which is the common multiplication, and not about the product of constructs like vectors and matrices. We can define mathematical operations on other entities, not just numbers, as long as it suits us. The product of a vector with a matrix is such an operation. There is no mystery involved in it: it is simply an operation that we define as a particular calculation involving the elements of the vector and matrix.

Suppose that we make bagels and croissants that we sell for $2.00 and $1.50, respectively. We have two shops; on a particular day, the first shop sells 10 bagels and 20 croissants, while the second shop sells 15 bagels and 10 croissants. How do we find the total sales per shop?

To find the total sales from the first shop, we will multiply the price of a bagel with the number of bagels sold in that shop, and the price of a croissant with the number of croissants sold there, and we'll add these two:

$$2.00 \times 10 + 1.50 \times 20 = 50$$

We'll do the same thing to find the total sales from the second shop:

$$2.00 \times 15 + 1.50 \times 10 = 45$$

To express this more succinctly, we write down the prices for the bagels and croissants as a vector:

$$\begin{bmatrix} 2.00 & 1.50 \end{bmatrix}$$

We also write down the daily sales in a matrix. The matrix will have two columns, one per shop, and two rows, one for the bagels and one for the croissants:

$$\begin{bmatrix} 10 & 15 \\ 20 & 10 \end{bmatrix}$$

Then to find the total sales per shop, we multiply the elements of the vector with each column of the sales matrix and add them up. This defines the product of the vector with the matrix:

$$\begin{aligned} &\begin{bmatrix} 2.00 & 1.50 \end{bmatrix} \times \begin{bmatrix} 10 & 15 \\ 20 & 10 \end{bmatrix} \\ &\quad = \begin{bmatrix} 2.00 \times 10 + 1.50 \times 20 & 2.00 \times 15 + 1.50 \times 10 \end{bmatrix} \\ &\quad = \begin{bmatrix} 50 & 45 \end{bmatrix} \end{aligned}$$

The product of a vector with a matrix is a special case of the product of two matrices. Let's extend the example so that instead of having a vector with the prices of the bagels and croissants, we have a matrix with the prices and profits per sale:

$$\begin{bmatrix} 2.00 & 1.50 \\ 0.20 & 0.10 \end{bmatrix}$$

To find the total sales per shop and total profit per shop, we will create a matrix in which the entries in the *i*th row and *j*th column will be the sum of products of the *i*th row of the prices and profits matrix with the *j*th row of the sales matrix. This is the definition of the product of the two matrices:

$$\begin{bmatrix} 2.00 & 1.50 \\ 0.10 & 0.20 \end{bmatrix} \times \begin{bmatrix} 10 & 15 \\ 20 & 10 \end{bmatrix}$$
$$= \begin{bmatrix} 2.00 \times 10 + 1.50 \times 20 & 2.00 \times 15 + 1.50 \times 10 \\ 0.10 \times 10 + 0.20 \times 20 & 0.10 \times 15 + 0.20 \times 10 \end{bmatrix}$$
$$= \begin{bmatrix} 50 & 45 \\ 5 & 3.5 \end{bmatrix}$$

Returning to pagerank, in each round the calculation of the pagerank vector is really the product of the value of the pagerank vector in the previous round with the hyperlink matrix. As we go through the rounds, we get successive estimates of the pageranks—that is, successive estimates of the pagerank vector that is made up of them.

To get these successive estimates of the pagerank vector we only need to multiply the vector in each round with the hyperlink matrix, thereby getting the vector for the next round.

In the first round, we start with a pagerank vector whose contents are all equal to $1 / n$, where n is the number of pages. If we denote this first pagerank vector by π_1, the pagerank vector at the end of the first round by π_2, and the hyperlink matrix by H, we have:

$$\pi_2 = \pi_1 \times H$$

In each round we use the pagerank vector of that round to calculate the pagerank vector for the following round. In the second voting round, where we got our third pagerank estimates—that is, our third pagerank vector—we performed the calculation:

$$\pi_3 = \pi_2 \times H = (\pi_1 \times H) \times H = \pi_1 \times (H \times H) = \pi_1 \times H^2$$

In the third voting round, we got our fourth pagerank vector:

$$\pi_4 = \pi_3 \times H = (\pi_1 \times H^2) \times H = \pi_1 \times (H^2 \times H) = \pi_1 \times H^3$$

As in every iteration, we multiply the result of the previous iteration by the hyperlink matrix, and in the end this is a series of products of the successive estimates of the pagerank vector by the hyperlink matrix. As we see, this is equivalent to multiplying the initial pagerank vector with increasing powers of the hyperlink matrix. This method of calculating successive approximations is called the *power method*. We see therefore that the calculation of the pageranks of a set of web pages is an application of the power method to the pagerank vector and hyperlink matrix, until the resulting pagerank vector does not change, or as we say, until it *converges* to a stable value—our final pagerank metrics.

We have just reached a more precise description of how to calculate the pageranks of a web graph:

1. Form the hyperlink matrix of the graph.

2. Start with initial pagerank estimates, giving a pagerank of 1 / n to each page, where n is the total number of pages.

3. Apply the power method, multiplying the pagerank vector by the hyperlink matrix until the values of the pagerank vector converge.

Apart from being succinct, this formulation allows us to transfer the problem to the realm of linear algebra, the branch of mathematics that treats matrices and operations on them. There is a well-established body of theory that we can use to investigate the power method as well as performant implementations of matrix operations, such as the multiplication that we described. The matrix formulation of the problem will also help investigate whether the power method will *always* converge so that we can always come up with a solution to the pageranks of a graph.

Dangling Nodes and the Random Surfer

We now turn to an example of a simpler graph, consisting of just three nodes:

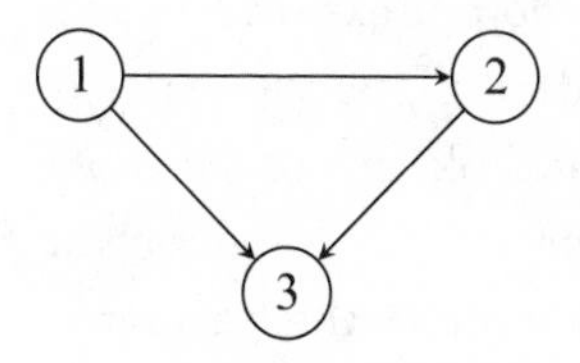

We want to find the pageranks of these three nodes. We follow the same algorithm. We initialize the pagerank vector to 1 / 3, giving equal pageranks to all nodes. Then we multiply the pagerank vector with the hyperlink matrix, which is:

$$\begin{bmatrix} 0 & 1/2 & 1/2 \\ 0 & 0 & 1 \\ 0 & 0 & 0 \end{bmatrix}$$

If we start the iterations of the power method, multiplying the pagerank vector with the hyperlink matrix to update the pagerank vector, and then again and again, we'll find out that after four iterations, all pageranks have gone down to zero:

Round	Page 1	Page 2	Page 3
start	0.33	0.33	0.33
1	0.00	0.17	0.50
2	0.00	0.00	0.17
3	0.00	0.00	0.00

That is clearly a problem. We do not expect all pages to have zero importance here. After all, page 3 has two backlinks and page 2 has one backlink, so somehow we would expect this to show on the results, let alone the fact that we also want the total sum of the pageranks to be one. Here nothing ended up being of any import at all.

The cause of the problem is node 3. Although this node has backlinks and would thereby gain importance, it has no outgoing links. So in a way it sucks importance from the rest of the graph, but does not redistribute it anywhere. It acts as a selfish node or black hole: what goes in, does not go out. After a few iterations, it has acted as a sink where all pagerank values have gone in and vanished.

Such nodes are called *dangling nodes* because they hang at the (dead) ends of the graph. On the web, nothing prohibits the existence of such pages. Although web pages usually have both incoming and outgoing links, a page with no outgoing links can appear and would wreak havoc with the power method as we have described it.

To overcome the problem, we work with a metaphor. We imagine that we have a human who surfs the web, jumping from page to page. To go from one page to another, the surfer normally follows a link. But then the surfer comes on a dangling node: a page with no links to any other page. We don't want our surfer to remain trapped in there so we give the surfer the capability to jump to any other page, anywhere on the web. It is as if we are surfing the web from page to page until we reach a dead end. When we get there, we don't give up and stop. We can always type another address in our web browser and move to any other web page we want, even if no links exist to it from the dangling page. This is what we want our surfer to do. When at a loss about where to go, the surfer will pick a page, any page, from the web and go there to continue surfing. The surfer becomes a *random surfer*, equipped with a teleportation device that can take the surfer instantly to any place at all.

To take this metaphor back to pagerank, we interpret the hyperlink matrix as giving us the probabilities that a surfer will follow a link to go to a particular page. In our three-nodes example, the first row of the hyperlink matrix tells us that when on page 1, the surfer will choose either page 2 or 3 with equal probability. The second row tells us that when on page 2, the surfer will always choose to visit page 3. Going back to our first example for a moment, if the surfer lands on page 5, then it is possible to go to page 2, 3, or 4 with a probability of 1 / 3 for each of these outcomes.

A dangling node manifests itself in the presence of a row full of zeros. Then there is no probability that the surfer will go anywhere. This is where the random surfer kicks in. As we said, that surfer will jump to any page in the graph. That means that in effect, we change the hyperlink matrix so that it no longer has rows with zeros. As we want the surfer to jump to any web page with equal probability, instead of zeros we'll fill the row with $1 / n$, or in our example, 1 / 3. Our matrix will become:

$$\begin{bmatrix} 0 & 1/2 & 1/2 \\ 0 & 0 & 1 \\ 1/3 & 1/3 & 1/3 \end{bmatrix}$$

Now the surfer who lands on page 3 can go to any page in the graph with equal probability. The surfer may even stay temporarily on the same page, but that does not matter, as the surfer can try again and again, and at some point a different target page will be selected at random.

We call this modified hyperlink matrix, where we change zero rows to rows with values equal to 1 / *n*, the *S* matrix. If we run the power method using the *S* matrix, then the evolution of the pageranks will be:

Round	Page 1	Page 2	Page 3
start	0.33	0.33	0.33
1	0.11	0.28	0.61
2	0.20	0.26	0.54
3	0.18	0.28	0.54
4	0.18	0.27	0.55
5	0.18	0.27	0.54

This time the algorithm converges to nonzero values; no sucking out of importance occurs. Also, the results make sense. The highest pagerank is achieved by page 3, which has two backlinks; then comes page 2, with one backlink, and then page 1, which has no backlinks at all.

The Google Matrix

We seem to have solved the problem, but a similar issue raises its head in more complex situations. The following graph has no dangling nodes:

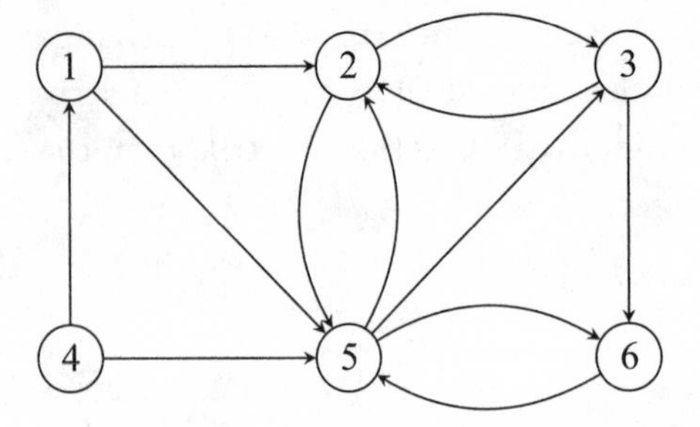

If we run the algorithm, we find that two nodes, pages 1 and 4, end up with zero pagerank:

Round	Page 1	Page 2	Page 3	Page 4	Page 5	Page 6
start	0.17	0.17	0.17	0.17	0.17	0.17
1	0.08	0.22	0.14	0.00	0.42	0.14
2	0.00	0.25	0.25	0.00	0.29	0.21
3	0.00	0.22	0.22	0.00	0.33	0.22

What happened is that even though there is no dangling node, there is a set of nodes that act as a sink for the rest of the graph. If you scrutinize the graph, you will see that the nodes 2, 3, 5, and 6, taken together as a group, have only incoming links. It is possible to go from node 1 or 4 to this group, but once we are in, we can only move inside the group. We are not able to go outside. Our random surfer will be trapped, not inside a single web page this time, but inside a group of pages that link only between themselves.

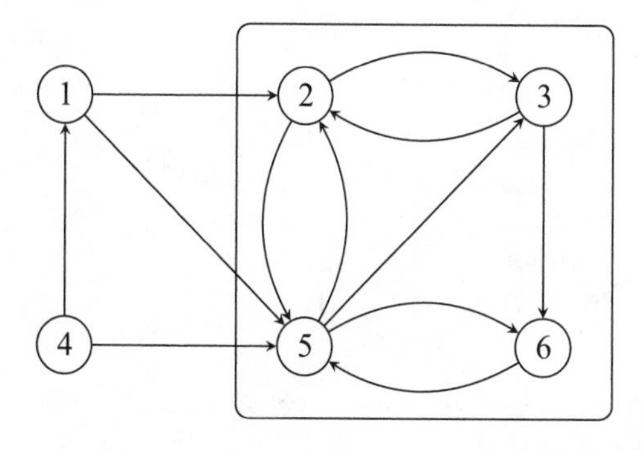

We need again to help the random surfer escape from this trap. This time the solution requires more comprehensive changes to the hyperlink matrix. Our initial hyperlink matrix allowed the surfer to go from page to page only using the existing links in the original graph. Then we modified the hyperlink matrix to handle rows with all zero elements and came up with the S matrix that allowed the surfer to get away from dangling nodes. This enabled the random surfer to jump to anywhere in the graph when in a dangling node. Now we will change the behavior of the random surfer a bit more by modifying the S matrix.

Right now, when a surfer lands on a node, the possible moves are those indicated by the S matrix. In the last example, the S matrix is the same as the hyperlink matrix because no zero rows exist:

$$\begin{bmatrix} 0 & 1/2 & 0 & 0 & 1/2 & 0 \\ 0 & 0 & 1/2 & 0 & 1/2 & 0 \\ 0 & 1/2 & 0 & 0 & 0 & 1/2 \\ 1/2 & 0 & 0 & 0 & 1/2 & 0 \\ 0 & 1/3 & 1/3 & 0 & 0 & 1/3 \\ 0 & 0 & 0 & 0 & 1 & 0 \end{bmatrix}$$

If the random surfer lands on page 5, then the possible moves are to pages 2, 3, or 6, all with 1 / 3 probability, as the S matrix indicates. We will make the random surfer more agile, with the power to move following the S matrix *not always*, but with some probability a that we will choose; then for some probability $(1 - a)$, the random surfer will jump anywhere in the graph, unconstrained by the S matrix.

The ability to jump from anywhere to anywhere in the graph means that we cannot have any zeros at all in the matrix—because a zero entry denotes a move that cannot be made. To achieve what we want, we will need to *increase* the zero entries in a row by some value and *decrease* the nonzero entries so that the whole row always sums up to one. The exact final values of the matrix can be calculated through linear algebra, based on S and the probability a.

The new matrix that will be derived is called the *Google matrix*, and we use the symbol G. If the behavior of the random surfer is determined by the Google matrix, it will work out as we want: the surfer will appear to be following the S matrix with probability a and move independently with probability $(1 - a)$. In our example, the Google matrix is:

$$\begin{bmatrix} \frac{3}{120} & \frac{54}{120} & \frac{3}{120} & \frac{3}{120} & \frac{54}{120} & \frac{3}{120} \\ \frac{3}{120} & \frac{3}{120} & \frac{54}{120} & \frac{3}{120} & \frac{54}{120} & \frac{3}{120} \\ \frac{3}{120} & \frac{54}{120} & \frac{3}{120} & \frac{3}{120} & \frac{3}{120} & \frac{54}{120} \\ \frac{54}{120} & \frac{3}{120} & \frac{3}{120} & \frac{54}{120} & \frac{3}{120} & \frac{3}{120} \\ \frac{3}{120} & \frac{37}{120} & \frac{37}{120} & \frac{3}{120} & \frac{3}{120} & \frac{37}{120} \\ \frac{3}{120} & \frac{3}{120} & \frac{3}{120} & \frac{3}{120} & \frac{105}{120} & \frac{3}{120} \end{bmatrix}$$

Compare that to the S matrix. Observe that in the first row, we had two entries with $1 / 2$ and the rest were zero. Now in the Google matrix, we have the two $1 / 2$ entries turned to $54 / 120$, and the rest of the entries turned from 0 to $3 / 120$. Similar transformations have occurred in the other rows. If, then, the random surfer lands on page 1, the possible moves out are to pages 2 and 5 with probability

54 / 120 for either of them, or any other page with probability 3 / 120 for each one of them.

We are now able to give the final definition of the PageRank algorithm:

1. Form the Google matrix of the graph.

2. Start with initial pagerank estimates, giving a pagerank of $1 / n$ to each page, where n is the total number of pages.

3. Apply the power method, multiplying the pagerank vector by the Google matrix until the values of the pagerank vector converge.

We simply substituted "Google matrix" for "hyperlink matrix" of the initial algorithm. If we trace this algorithm in our graph with the group of sink nodes, we'll get:

Round	Page 1	Page 2	Page 3	Page 4	Page 5	Page 6
start	0.17	0.17	0.17	0.17	0.17	0.17
1	0.10	0.14	0.14	0.10	0.31	0.21
2	0.07	0.15	0.17	0.07	0.31	0.23
3	0.05	0.14	0.18	0.05	0.32	0.26
4	0.05	0.14	0.17	0.05	0.33	0.27

It works out fine; we get no zero pageranks anymore.

The power method with the Google matrix will work always. Linear algebra tells us that it will always converge to a final set of pagerank values, the sum of which will be one, without suffering from dangling nodes or parts of the graph draining the pageranks of the rest of the graph. We don't even need to initialize the pageranks to exactly $1 / n$ when we start. Any initial set of values will do, as long as they sum up to one.

PageRank in Practice

Having established that we have a method to find the pageranks in any graph, the question remains whether the results are in the end sensible.

The pagerank vector, in the way that we have defined it, is a special vector in relation to the Google matrix. When the power method finishes, the pagerank vector does not change any more. Therefore if we multiply the Google matrix by the pagerank vector we will get simply the same pagerank vector. In linear algebra, this vector is called the *first eigenvector* of the Google matrix. Without going deep into the mathematics, the underlying theory supports the notion that this vector has some special significance to the matrix.

Beyond mathematics, the final arbiter of whether PageRank is a good way to assign importance to web pages is the utility of its results to us humans. The Google search engine gives good results, meaning that the results are in accordance with what we, the users of the search engine, regard as being important. If the pagerank vector was a mathematical curiosity that bore no relation to the significance of web pages, we would not be concerned with it today.

An additional advantage of PageRank is that it can be implemented efficiently. The Google matrix is huge; we need one row and one column for every single page on the web. Yet the Google matrix is derived, as we saw, from the S matrix, which in turn is derived from the hyperlink matrix. We do not really need to create and store the Google matrix itself; we can create it dynamically with matrix operations on the hyperlink matrix. This is convenient. In contrast to the Google matrix, which has no zeros anywhere, the hyperlink matrix has lots and lots of zeros. The web may have billions of pages, but every single page links to only a small number of other web pages. The hyperlink matrix is what we call a *sparse matrix*: one that is mostly full of zeros, with only some nonzero entries, which are scales of magnitude fewer than the zero entries. Thus we can store the matrix using clever techniques that instead of requiring a big slab of memory to fill with mostly zeros and a few nonzeros, store only the positions where the nonzeros occur. Rather than storing the whole hyperlink matrix, we need only store the coordinates of the nonzero entries, which will require only a fraction of the storage space. This gives us big leverage in the practical implementations of the PageRank algorithm.

Finally, an important caveat. Although we know that PageRank played a crucial role in the success of Google, we do not know how, or even if, PageRank is used in Google today. The Google search engine has been evolving during the years, and the changes are not made public. We know that Google uses our past searches to fine-tune the results that it presents to our queries. It can tune the results depending on the country that we live in. It can also take into account the overall trends in the queries that other people make all around the world. All these are part of the secret sauce that Google uses to improve its product and retain its position in the search engine business against competitors. This, however, does not detract from the algorithm's efficiency in solving the problem of ranking web pages, represented as nodes in a graph.[1]

PageRank highlights an additional aspect of algorithms. The success of an algorithm does not hinge only on its elegance and efficiency. It also has to do with the mapping of the algorithm to a problem. This is a creative act. To solve the problem of web search, one has to overcome the issue of the sheer size of the web. But once you conceive of the web as a graph, its size turns into an advantage, not a hindrance. It is exactly because there are so many pages, hyperlinked to each other, that you may expect that a method that is based on the link structure of the graph will in the end work. Finding the way to model a problem is the first step in finding the way to solve it with an algorithm.

6

DEEP LEARNING

Deep learning systems have burst onto the scene in recent years, often making headlines in mainstream media. There we see computer systems performing feats that were the purview of humans. Even more tantalizing is the fact that these systems are frequently presented as having some similarities to the way the human mind works—which of course cues to the idea that perhaps the key for artificial intelligence may be to mimic the workings of human intelligence.

Brushing aside the hype, most scientists working on deep learning do not ascribe to the view that deep learning systems work like the human mind. The goal is to exhibit some useful behavior, which we often associate with intelligence. We do not go about copying nature, however; in fact, the architecture of the human brain is much too complicated to emulate on a computer. But we do take some leaves out of nature's book, simplify them a lot, and try to engineer systems that could, in certain fields, do things usually done by biological systems that have evolved over millions of years. Moreover, and this concerns us here in this book, deep learning systems can be understood in terms of the algorithms they employ. This will shed some light on what they do exactly, and how. And it should help us see that underneath their accomplishments, the main ideas are not complicated. That should not belittle the achievements of the field. We'll see that deep learning requires an enormous amount of human ingenuity in order to come to fruition.

To understand what deep learning is about, we need to start small, from humble beginnings. On these we will build a more and more elaborate picture, until, at the end of the chapter, we will be able to make sense of what the "deep" in deep learning stands for.

Neurons, Real and Artificial

Our starting point will be the main building block of deep learning systems, which does come from biology. The brain is part of the nervous system, and the main components of the nervous system are cells called *neurons*. Neurons have a particular shape; they look different from the globular structures that we usually associate with cells. You can see below one of the first images of neurons, drawn in 1899 by the Spanish Santiago Ramón y Cajal, a founder of modern neuroscience.[1]

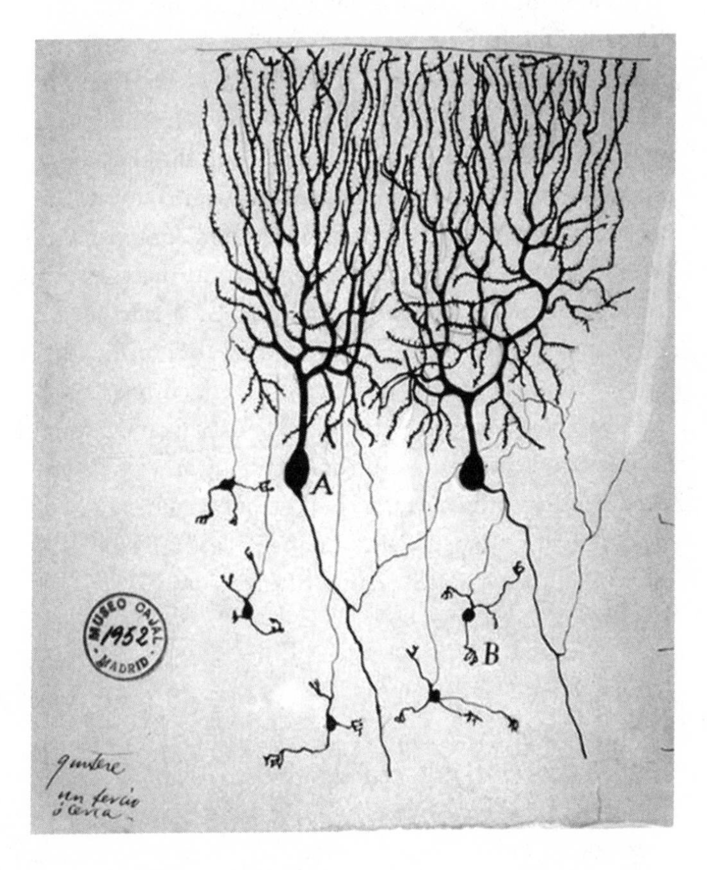

The two structures that stand out in the middle of the image are two neurons of the pigeon brain. As you can see, a neuron consists of a cell body and the filaments that extrude from it. These filaments connect a neuron to other neurons through *synapses*, embedding the neurons in a network. The neurons are asymmetrical. There are many filaments on the one side and one filament on the other side of each neuron. We can think of the many filaments on the one side as the neuron's inputs, and the long outgoing filament on the other side as the neuron's output. The neuron takes input in the form of electric signals from its incoming synapses and may send a signal to other neurons. The more inputs it receives, the more likely it is to output a signal. We say that the neuron then *fires* or is *activated*.

The human brain is a vast network of neurons, which number about one hundred billion, and each one of them is connected on average to thousands of other neurons. We do not have the means to build anything like that, but we can build systems out of simplified, idealized models of neurons. This is a model of an artificial neuron:

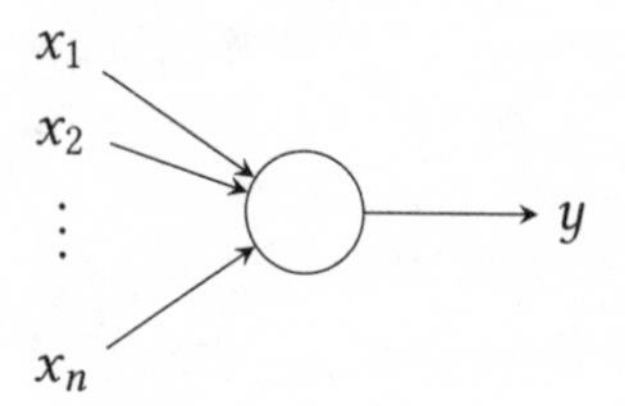

That is an abstract version of a biological neuron, being just a structure with a number of inputs and one output. The output of a biological neuron depends on its input; similarly, we want the artificial neuron to be activated depending on its input. We are not in the realm of brain biochemistry, but in the world of computing, so we need a computational model for our artificial neuron. We assume that the signals received and sent by neurons are numbers. Then the artificial neuron takes all its inputs, calculates some arithmetic value based on them, and produces some result on its output. We do not need any special circuit for implementing an artificial neuron. You can think of it as a small program inside a computer that takes its inputs and produces an output, much like any other computer program. We do not need to build artificial neural networks literally; we can and do simulate them.

Part of the learning process in biological neural networks is the strengthening or weakening of the synapses between neurons. The acquisition of new cognitive abilities and absorption of knowledge result in some synapses between neurons getting stronger, while others get weaker or even drop off completely. Moreover, synapses may not only excite a neuron to fire but also inhibit its activation; when a signal arrives on that synapse, the neuron should not fire. Babies have actually more synapses in their brains than adults. Part of growing up is pruning the neural network inside our heads. Perhaps we could think of the infant brain as a block of marble; as we go through the years in our lives, the block is chipped through our experiences and the things we learn, and a form emerges.

In an artificial neuron, we approximate the plasticity of synapses, their excitatory or inhibitory role, through *weights* we apply to the inputs. In our model artificial neuron, we have n inputs, $x_1, x_2, \ldots, x_n$. To each one of them we apply a weight, $w_1, w_2, \ldots, w_n$. Each weight is multiplied by the corresponding input. That final input received by a neuron is the sum of the products: $w_1 x_1 + w_2 x_2 + \cdots + w_n x_n$. To this *weighted input* we add a *bias* b, which you can think of as the propensity the neuron has to fire; the higher the bias, the more likely it is to be activated, while a negative bias added to the weighted input will actually inhibit the neuron from firing.

The weights and bias are the *parameters* of the neuron because they influence its behavior. As the output of a biological neuron depends on its inputs, so the output of an artificial neuron depends on the input it gets. This happens by feeding the input into a special *activation function*, the result of which is the neuron's output. This is what happens, diagrammatically, using $f(\cdot)$ as a stand-in for the activation function:

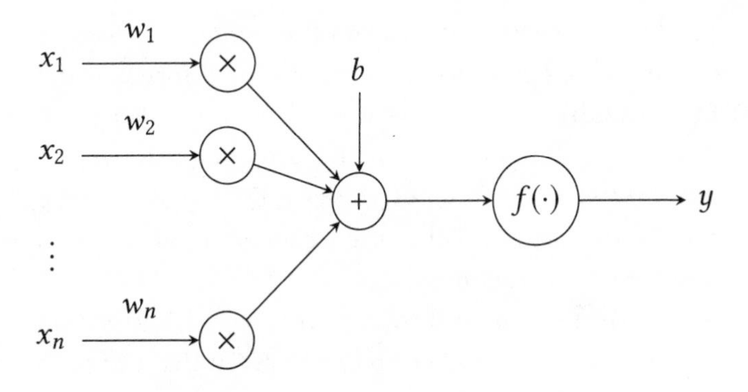

The simplest activation function is a step function, giving us a result of 0 or 1. The neuron fires and outputs 1 if the input to the activation function is greater than 0, or stays silent outputting 0 otherwise:

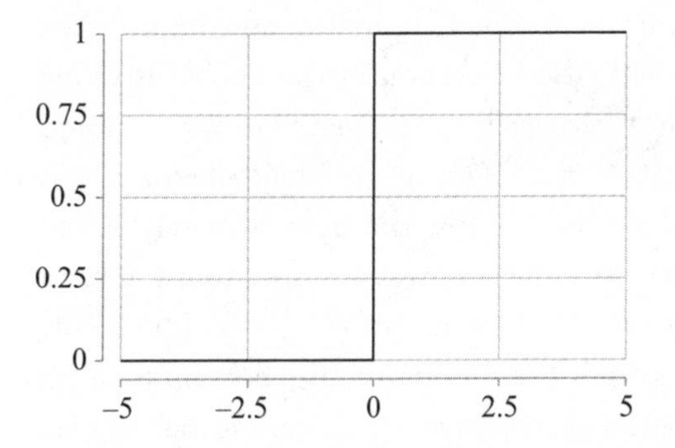

Instead of a bias, it is helpful to think of a threshold. The neuron outputs 1 if the weighted input exceeds a threshold or outputs 0 otherwise. Indeed, if we write the behavior of the neuron as a formula, the first condition is $w_1x_1 + w_2x_2 + \cdots + w_nx_n + b > 0$ or $w_1x_1 + w_2x_2 + \cdots + w_nx_n > -b$. By using $t = -b$, we get $w_1x_1 + w_2x_2 + \cdots + w_nx_n > t$, where t, the opposite of the bias, is the threshold that the weighted input needs to pass for the neuron to fire.

In practice we tend to use other, related activation functions instead of the step function. On the next page you can see three common ones.

The one on the top is called *sigmoid* because it has an S shape.[2] Its output ranges from 0 to 1. A large positive input results in outputs close to 1; a large negative input results in an output close to 0. This approximates a biological neuron that fires on large inputs and stays silent otherwise, and is a smooth approximation to the step function. The activation function in the middle is called *tanh*, short for *hyperbolic tangent* (there are various ways to pronounce it: "tan-H," "then," or "thents" with a soft th, as in thanks).[3] It looks like the sigmoid function, but it differs in that its output ranges from −1 to +1; a large negative input results in a negative output, mimicking an inhibitory signal. The function at the bottom is called a *rectifier*; it turns all negative inputs to 0, otherwise its output is directly proportional to its input. The following table shows the output of the three activation functions for different inputs.

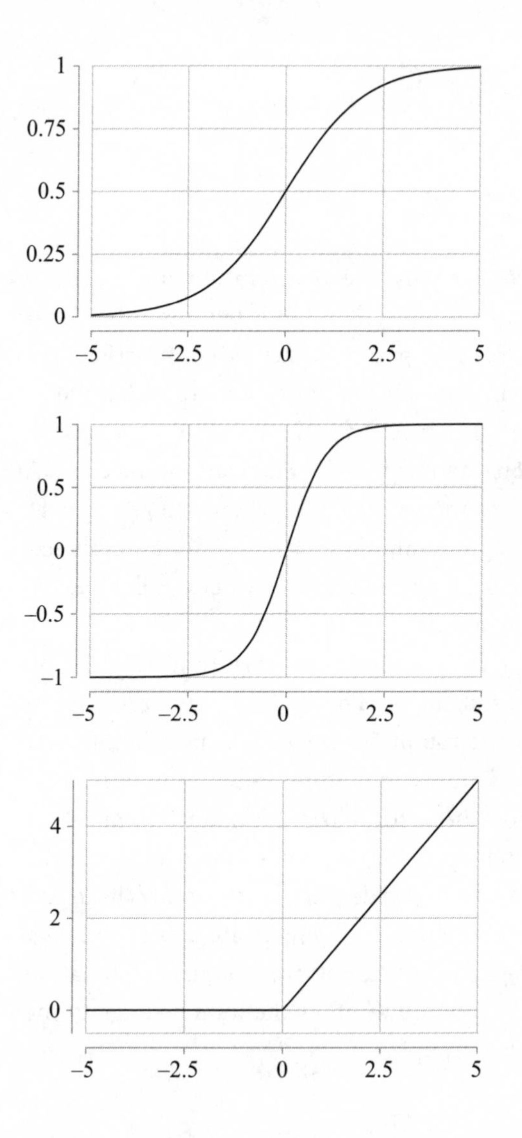
1
0.75
0.5
0.25
0
–5
–2.5
0
2.5
5
1
0.5
0
–0.5
–1
–5
–2.5
0
2.5
5
4
2
0
–5
–2.5
0
2.5
5

	-5	-1	0	1	5
sigmoid	0.01	0.27	0.5	0.73	0.99
tanh	-1	-0.76	0	0.76	+1
rectifier	0	0	0	1	5

If you wonder why the proliferation of activation functions (there are also others), it is because it has been found in practice that particular activation functions are more suitable in some applications than others. As the activation function is crucial for the behavior of a neuron, neurons are often named by their activation functions. A neuron that uses the step function is called a *Perceptron*.[4] Then we have sigmoid and tanh neurons. We also call neurons *units*, and a neuron using the rectifier is called a *ReLU*, for rectified linear unit.

A single artificial neuron can learn to distinguish between two sets of things. For example, take the data in the figure on the top of the next page, portraying a set of observations with two features, x_1, on the horizontal axis, and x_2, on the vertical axis. We want to build a system that will tell apart the two blobs. Given any item, the system will be able to decide whether the item falls in one group or another. In effect, it will create a *decision boundary*, like in the figure at the bottom. For any combination of (x_1, x_2), it will tell us whether the item belongs to the lighter or darker group.

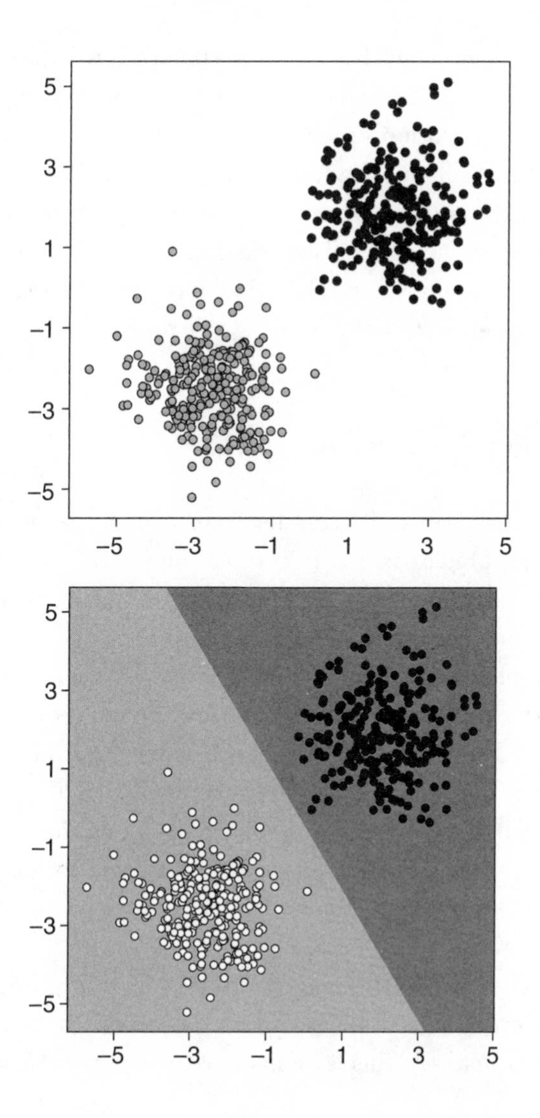
5
3
1
−1
−3
−5
−5
−3
−1
1
3
5
5
3
1
−1
−3
−5
−5
−3
−1
1
3
5

The neuron will have only two inputs. It will take each (x_1, x_2) pair and calculate an output. If we are using the sigmoid activation function, the output will be between 0 and 1. We'll take the values greater than 0.5 to fall into one group and the other values to fall into the other. In this way the neuron will act as a *classifier*, sorting our data into distinct classes. But how does it do that? How can the neuron get to the point of being able to classify data?

The Learning Process

At the moment of its creation, our neuron cannot recognize any kind of data; it *learns* to recognize them. The way it learns is by example. The whole process is akin to having a student learn something by giving them a large bunch of problems on a subject, along with their solutions. We ask the student to study each problem and its solution. If they are diligent, we expect that after the student has gone through a number of problems, they will have figured out how to get from a problem to its solution and will even be able to solve new problems, related to the ones they studied, but this time without having recourse to any solutions.

When we do this, we *train* the computer to find the solutions; the set of solved example problems is called the *training data set*. This is an instance of *supervised learning*

At the moment of its creation, our neuron cannot recognize any kind of data; it *learns* to recognize them. The way it learns is by example.

because the solutions guide the computer, like a supervisor, toward finding the right answers. Supervised learning is the most common form of *machine learning*, the entire discipline that deals with methods where we train computers to do things. Apart from supervised learning, machine learning also encompasses *unsupervised learning*, where we provide the computer with a training data set, but not with any accompanying solutions. There are important applications of unsupervised learning, like, for example, grouping observations into different clusters (there is no a priori solution to what a correct cluster of observations is). In general, though, supervised learning is more powerful than unsupervised learning, as we provide more information during training. We will only deal with supervised learning here.

After training, the student often passes some tests to see how well they mastered the material. Similarly, in machine learning, after training we give the computer another data set that it has not seen before and ask it to solve this *test data set*. Then we evaluate the performance of the machine learning system based on how well it manages to solve the problems in the test data set.

In the classification task, training for supervised learning works by giving the neuron network a large number of observations (problems) along with their classes (solutions). We expect that the neuron will somehow learn how to get from an observation to its class. Then if we give it an observation it has not seen before, it should classify it with reasonable success.

The behavior of a neuron for any input is determined by its weights and bias. When we start, we set them at random values; the neuron knows nothing, like a clueless student. We give the neuron one input in the form of a (x_1, x_2) pair. The neuron will produce an output. As we have random weights and bias, the output will also be random. For each of our observations in the training data set, however, we do know what the correct answer from the neuron should be. We can then calculate how far off the neuron's output is from the desired one. This is called the *loss*: a measure of how wrong the neuron is for a given input.

For example, if for an input the neuron produces as output the value 0.2, while the desired output is 1.0, we can calculate the loss by the difference between the two values. To avoid having to deal with signs, we usually take as the loss the square of the difference; here it would be $(1.0 - 0.2)^2 = 0.64$. If the desired output were 0.0, then the loss would be $(0.0 - 0.2)^2 = 0.04$. Be it as it may, having calculated the loss, we can now adjust the weights and bias so as to minimize it.

Going back to the human student, after each failed attempt to solve an exercise, we nudge them to perform better. The student figures out that they have to change their approach a bit and try with the next example. If they fail, we nudge them again. And again. Until after a lot of examples in the training data set, they will start getting things right more and more, and will be able to tackle the test data set.

When a student learns, neuroscience tells us that the wiring inside the brain changes; some synapses between neurons get stronger, some get weaker, and some are dropped. There is no direct equivalent to an artificial neuron, but something similar happens. Recall once more that the behavior of a neuron depends on its input, weights, and bias. We have no control over the input; it comes from the environment. But we can change the weights and biases. And this is what really happens. We update the values of the weights and bias in such a way that the neuron will minimize its errors.

The way that the neuron achieves that is by taking advantage of the nature of the task it is called to perform. We want it to take each observation, calculate an output corresponding to a class, and adjust its weights and bias to minimize its loss. So the neuron is trying to solve a *minimization* problem. Given an input and the output it produces, the problem is, *How are we to recalibrate the weights and bias to minimize the loss*?

This requires a conceptual change of focus. Up to this point we have described a neuron as something that takes some inputs and produces an output. Viewed in this way, the whole neuron is a big function that takes its inputs, applies the weights, sums the products, adds the bias, passes the result through the activation function, and produces the final output. But if we think of it another way, our inputs and outputs are actually given (that is our training data set), while what we can change are the weights and bias. So we can view the whole neuron as a function whose variables are *the weights and bias* because these are what we can really affect, and for every input we want to change them so as to minimize the loss.

If we take as an illustration a simple neuron, with just one weight and no bias, then the relationship between the loss and weight might be as in the left part of the figure on the next page. The thick curve shows the loss as a function of the weight for a given input. The neuron should adjust its weight so that it reaches the minimum value of the function. The neuron, for the given input, has currently a loss at the indicated point. Unfortunately, the neuron does not know what is the ideal weight that would minimize the loss, given that the only thing it does know is the value of the function at the indicated point; it is not endowed with a vantage point of view like we have with the figure at our disposal. The neuron may only adjust its weight by a small amount—either increase or decrease it—so that it moves closer to the minimum.

To find out what to do, whether to increase or decrease the weight, the neuron can find the tangent line at the current point. Then it can calculate the slope of the tangent line; this is the angle with the horizontal axis, which we have also shown in the figure. Note that the neuron can do that without any special capabilities apart from being able to carry out calculations at the local point. The slope of the tangent is negative because the angle is clockwise. The slope shows the *rate of change of a function*; therefore a negative slope indicates that by increasing the weight, the loss decreases. The neuron thereby discovers that to decrease the loss, it has to move to the right. As the slope is negative and the required change in the weight is positive, the neuron finds that it must move the weight in a positive direction—opposite to what is indicated by the slope.

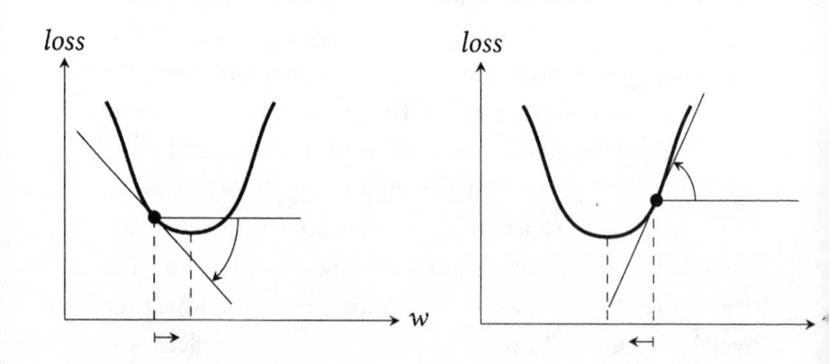

Now turn to the figure on the right. This time the neuron is to the right of the minimum loss. It takes the tangent again and calculates its slope. The angle and therefore slope is positive. A positive slope indicates that by increasing the weight, the loss increases. The neuron then knows that in order to minimize the loss, it has to decrease the weight. As the slope is positive and the required change in the weight is negative, the neuron finds again that it must move in the opposite direction than that indicated by the slope.

In both cases, then, the rule is the same: the neuron calculates the slope and updates the weight in the opposite direction from the slope. All this might look familiar from calculus. The slope of a function at a point is its *derivative*. To decrease the loss, we need to change the weight by a small amount that is opposite to the derivative of the loss.

Now a neuron does not usually have a single weight but rather has many, and also has a bias. To find out how to adjust each individual weight and the bias, the neuron proceeds like we described for the single weight. In mathematical terms, it calculates the so-called *partial derivative* of the loss with respect to each individual weight and bias. For n weights and a bias, that will be $n+1$ partial derivatives in total. A vector containing all the partial derivatives of a function is called its *gradient*. The gradient is the equivalent of the slope when we have multivariable functions; it shows the direction along which we have to move to increase the value of the function. To decrease it, we move in the opposite direction. Thus to decrease the loss, the neuron updates each weight and the bias in the opposite direction than the one indicated by the partial derivatives forming its gradient.[5]

The calculations are not really performed by drawing tangents and measuring angles. There are efficient ways to find the partial derivatives and gradient, but we don't need to get into the details. What is important is that we have a well-defined way to adjust the weights and bias to improve the results of the neuron. With this at hand, the learning process can be described by the following algorithm:

For each input and desired output in the training data set,

1. Calculate the output of the neuron and loss.
2. Update the weights and bias of the neuron to minimize the loss.

Once we have completed a training by going through all the data in the training data set, we say that we have completed an *epoch*. Usually we do not leave it at this. We repeat the whole process for a number of epochs; it is as if the student, after going through all the study material, started all over again. We expect that the next time they'll do better, as this time they do not start from zero—they are not completely clueless—having already learned something from the previous epoch.

The more we repeat the training by adding epochs into our training regime, the better we get with the training data. But too much training can be a bad thing. A student who studies again and again the same set of problems will probably learn to solve them by rote—without really knowing how to solve any other problems that they have not encountered before. We see that happening when a seemingly well-prepared student fails abysmally in the exams. In machine learning, when we train the computer on a training data set, we say that it *fits* the data. Too much training results in what is called *overfitting*: excellent performance with the training data set, and bad performance with the test data set.

It can be proven that following this algorithm, a neuron can learn to classify any data that are *linearly separable*. If our data have two dimensions (like our example), then that means that they should be separable by a straight line. If our data have more features, not just (x_1, x_2), the principle is generalized. For three dimensions—that is, three inputs (x_1, x_2, x_3)—the data are linearly separable if they can be separated by a simple plane in the three-dimensional space. For more dimensions, we call the equivalent of the line and plane a *hyperplane*.

At the end of the training, our neuron has learned to separate the data. "Learned" means that it has found the right weights and bias, in the way we described: it started out with random values and then gradually updated them, minimizing the loss. Recall the figure with the two blobs, which the neuron learned to separate with a decision boundary. We got from the neuron below at the left, to the neuron at the right, where you can see the final values of its parameters.

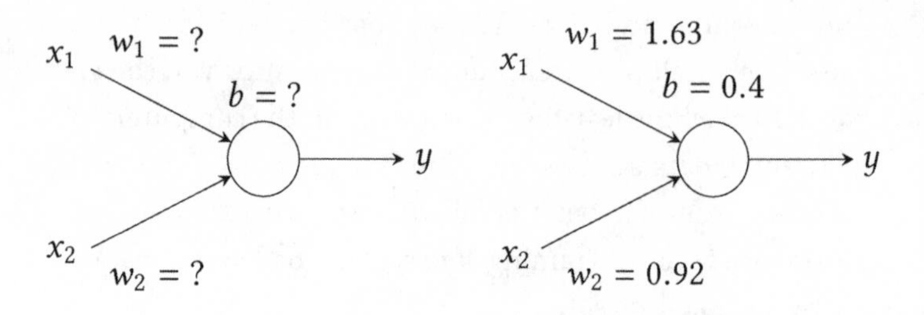

That does not always happen. A single neuron, acting alone, can only perform certain tasks, like this classification of linearly separable data. To handle more complicated tasks, we need to move from a lone artificial neuron to networks of neurons.

From Neurons to Neural Networks

As in biological neural networks, we can build *artificial neural networks* out of interconnected neurons. The input signals of a neuron can be connected to the outputs of other neurons, and its output signal can be connected to the inputs of other neurons. In this way we can create neural networks like this one:

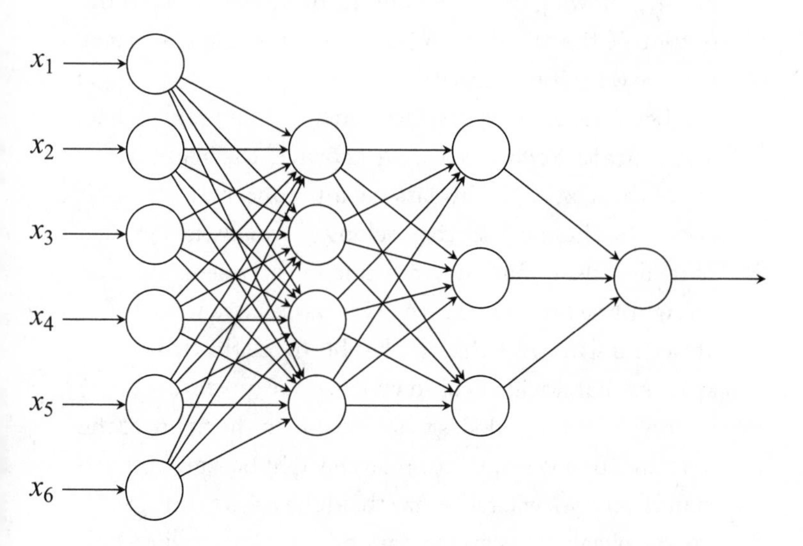

This artificial neural network has its neurons arranged in layers. This is often done in practice: many neural networks that we construct are made of layers of neurons, with each layer stacked next to a previous one. We have also made all the neurons on one layer connect to all the neurons on the next layer, going from left to right. This, again, is common, although not necessary. When we have layers connected like that, we call them *densely connected*.

While the first layer is not connected to any previous one, the output of the last layer is similarly not connected to any following layer. The output of the last layer is the output of the whole network; it will provide the values that we want it to calculate.

Let us return to a classification task. Our problem now is to pick apart two sets of data, shown in the figure on the top of the next page. The data fall into concentric circles. It is clear to a human that they belong to two distinct groups. It is also clear that they are not linearly separable: no straight line can separate the two classes. We want to create a neural network that will be able to tell the two groups apart so that it will tell us in which group any future observation will belong. This is what you see in the figure at the bottom. For any observation on the light background, the neural network will recognize that it belongs to one group; for any observation on the dark background, it will tell us that it belongs to the other group.

To achieve the results that we see in the lower figure, we build a network layer by layer. We put two neurons on the input layer, one for each coordinate of our data. We add one layer with four neurons, densely connected to the input layer. Because this layer is not connected to the input or output, it is a *hidden layer*. We add another hidden layer with two neurons, densely connected to the first hidden layer. We finish the network with an output layer of one neuron, densely connected to the last hidden layer. All

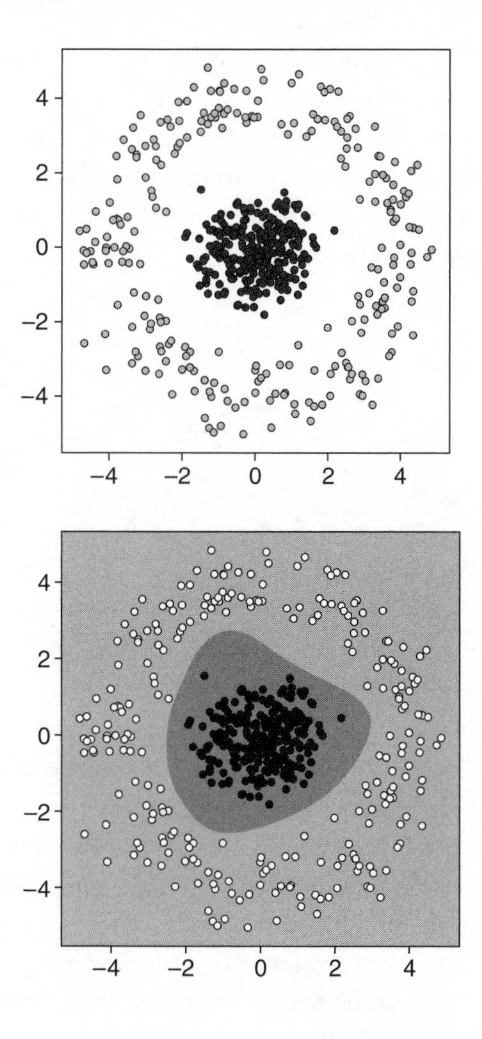
4
2
0
−2
−4
−4
−2
0
2
4
4
2
0
−2
−4
−4
−2
0
2
4

the neurons use the tanh activation function. The output neuron will produce a value between −1 and +1, displaying its belief that the data fall in one or the other group. We'll take that value and turn it into a binary decision, yes or no, depending on whether it exceeds 0.0 or not. This is what the neural network looks like:

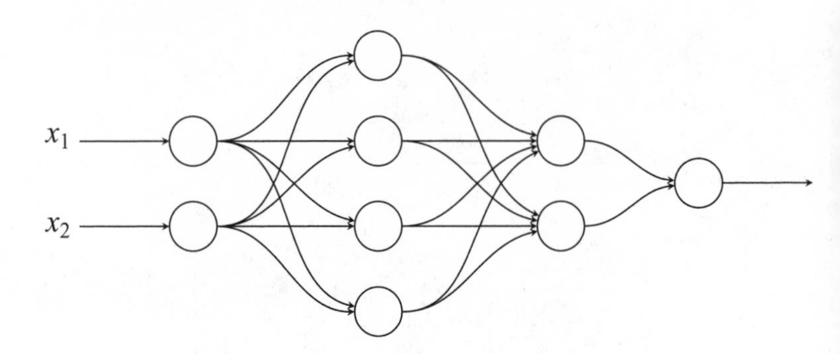

The Backpropagation Algorithm

In the beginning, the neural network knows nothing, and no adjustment has taken place; we start with random weights and biases. This is what ignorance means in the neural network world. Then we give the neural network an observation from our data—that is, a set of coordinates. The x_1 and x_2 coordinates will go on the input layer. Both neurons take the x_1 and x_2 values and they pass them as their output to the first hidden layer. All four neurons of that layer calculate their output, which in their turn, they send to the second hidden layer. The neurons on that layer send their own output to the neuron on the output layer, which produces the final output value of the neural network. As the calculations proceed from layer to layer, the neural network propagates the results of the neurons forward, from the input to the output layer:

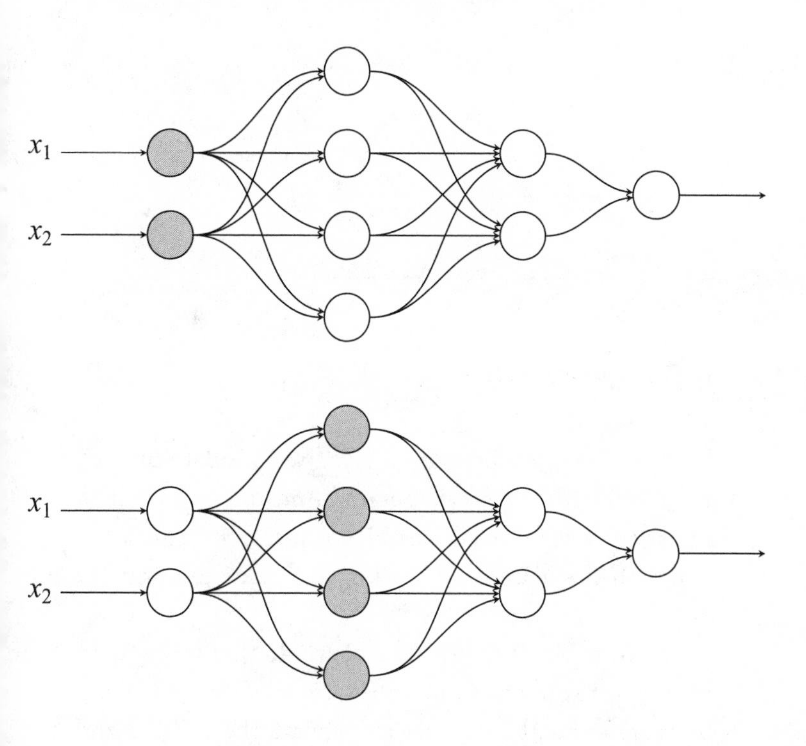

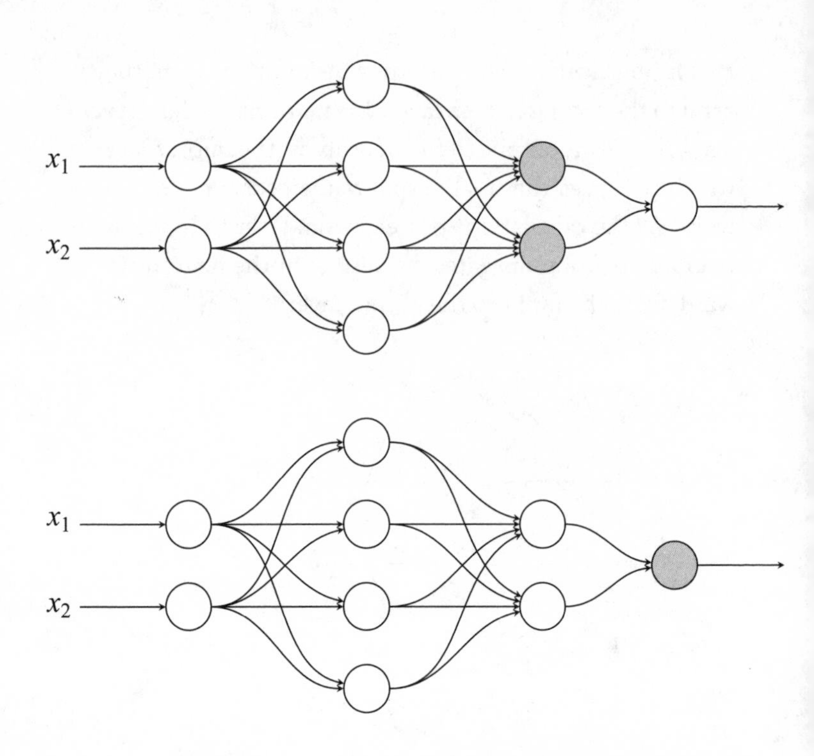

Once we reach the output layer, we calculate the loss, as we did with the single neuron. And then we want to adjust the weights and bias of not just one neuron but rather all the neurons in the network so as to minimize the loss.

It turns out that it is possible to do that by going in the opposite direction, from the output to the input layer. Once we know the loss, we can update the weights and biases of the neurons on the output layer (here we have just a single neuron, but this need is not always so). Having updated the neurons on the output layer, we can update the weights and biases of the neurons on the layer before that—the last hidden layer. Having done that, we can update the weights and biases of the layer before that—the one-but-last hidden layer. And so on, until we reach the input layer:

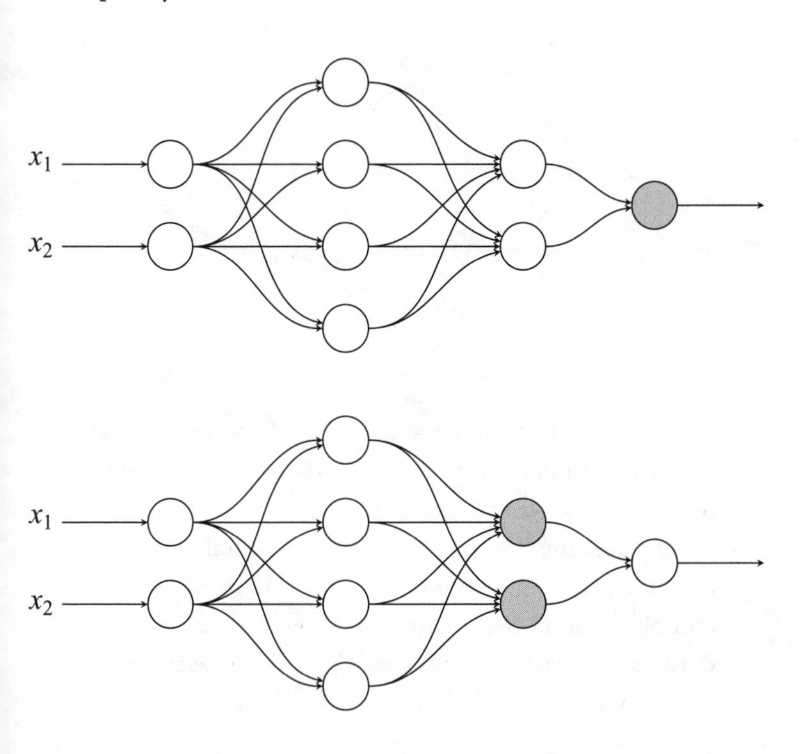

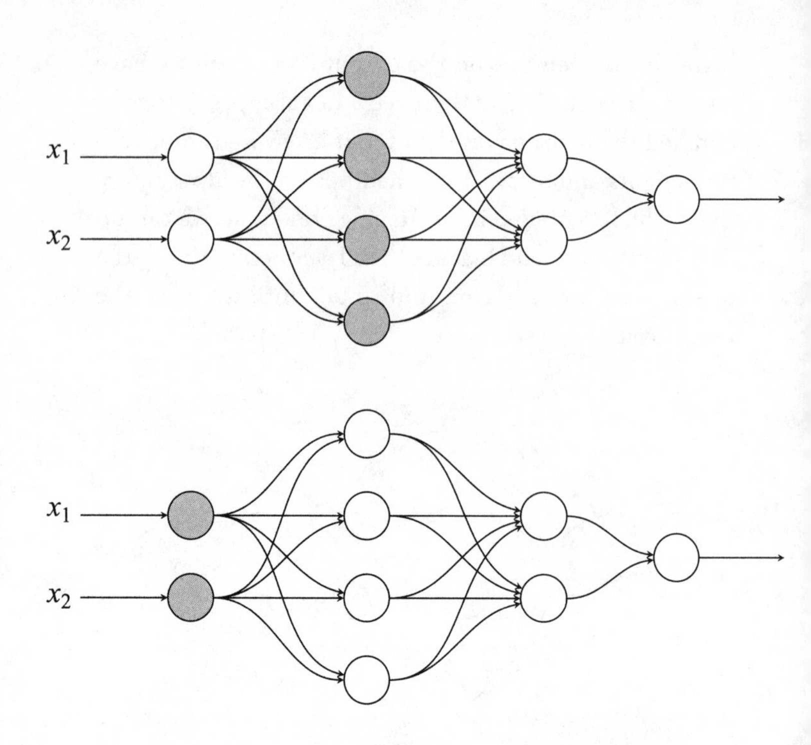

The way the weights and biases of the neurons are updated is similar to the way a single neuron is updated. Again, the updates are calculated based on mathematical derivatives. You can think of the whole neural network as an enormous function whose variables are the weights and biases of all the neurons. Then we can calculate the derivative of each and every weight and bias with respect to the loss, and use that derivative to update the neuron.

With this we arrive at the heart of the learning process in neural networks: the *backpropagation algorithm*.[6]

For each input and desired output,

1. Calculate the output and loss of the neural network proceeding layer by layer, going forward from the input to the output layer.
2. Update the weights and biases of the neurons to minimize the loss, going backward from the output to the input layer.

Using the backpropagation algorithm, we can build complex neural networks and train them to perform different tasks. The building blocks of deep learning systems are simple. They are artificial neurons, with their limited computational capabilities: taking inputs, multiplying by weights, summing, adding a bias, and applying an activation function on the resulting value. Their power derives from connecting lots and lots of them in special ways, where the resulting networks can be trained to perform the task that we want them to perform.

Recognizing Clothes

To render the discussion more concrete, let us assume that we want to build a neural network that recognizes items of clothing displayed in images, so this is going to be an *image recognition* task. Neural networks have been found to be exceptionally good at this.

Each image will be a small photo, of dimensions 28×28. Our training data set consists of 60,000 images, and our test data set consists of 10,000 images; we'll use 60,000 images for training the neural network, and another 10,000 images for evaluating how well it learned. Here is an example image, on which we have added axes and a grid to help the discussion that follows:[7]

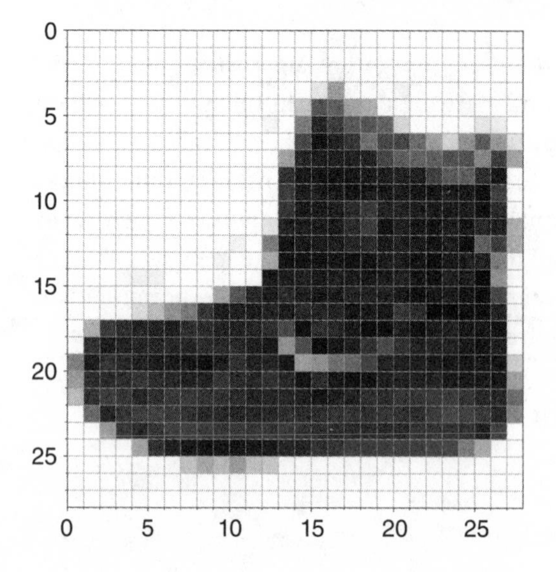

The image is broken into small distinct parts because that is how we handle images digitally. Taking the whole image as a rectangular plot, we divide it into small patches, $28 \times 28 = 784$ of them, and each patch is given an integer value from 0 to 255, corresponding to a shade of gray, with 0 being completely white and 255 being completely black. The above image is actually the matrix on the following page.

In reality, neural networks require that we usually scale their inputs to a small range of values, such as between 0 and 1, otherwise they may not work well; you may think of it as having large input values that lead neurons astray. That means that before using this matrix we would divide each cell by 255, but we'll ignore this in the rest of the discussion.

The different items of clothing may belong to ten different classes, which you can see in the table below. To a computer, the classes are just different numbers, which we call *labels*:

Label	Class	Label	Class
0	T-shirt/top	5	Sandal
1	Trouser	6	Shirt
2	Pullover	7	Sneaker
3	Dress	8	Bag
4	Coat	9	Ankle boot

0	0	0	0	0	0	0	0	0	0	0	0	0	0	0	0	0	0	0	0	0	0	0	0	0	0	0	0
0	0	0	0	0	0	0	0	0	0	0	0	0	0	0	0	0	0	0	0	0	0	0	0	0	0	0	0
0	0	0	0	0	0	0	0	0	0	0	0	0	0	0	0	0	0	0	0	0	0	0	0	0	0	0	0
0	0	0	0	0	0	0	0	0	0	0	0	1	0	0	13	73	0	0	1	4	0	0	0	0	1	1	0
0	0	0	0	0	0	0	0	0	0	0	0	3	0	36	136	127	62	54	0	0	0	1	3	4	0	0	3
0	0	0	0	0	0	0	0	0	0	0	0	6	0	102	204	176	134	144	123	23	0	0	0	0	12	10	0
0	0	0	0	0	0	0	0	0	0	0	0	0	0	155	236	207	178	107	156	161	109	64	23	77	130	72	15
0	0	0	0	0	0	0	0	0	0	0	1	0	69	207	223	218	216	216	163	127	121	122	146	141	88	172	66
0	0	0	0	0	0	0	0	0	1	1	1	0	200	232	232	233	229	223	223	215	213	164	127	123	196	229	0
0	0	0	0	0	0	0	0	0	0	0	0	0	183	225	216	223	228	235	227	224	222	224	221	223	245	173	0
0	0	0	0	0	0	0	0	0	0	0	0	0	193	228	218	213	198	180	212	210	211	213	223	220	243	202	0
0	0	0	0	0	0	0	0	0	1	3	0	12	219	220	212	218	192	169	227	208	218	224	212	226	197	209	52
0	0	0	0	0	0	0	0	0	0	6	0	99	244	222	220	218	203	198	221	215	213	222	220	245	119	167	56
0	0	0	0	0	0	0	0	0	4	0	0	55	236	228	230	228	240	232	213	218	223	234	217	217	209	92	0
0	0	1	4	6	7	2	0	0	0	0	0	237	226	217	223	222	219	222	221	216	223	229	215	218	255	77	0
0	3	0	0	0	0	0	0	0	62	145	204	228	207	213	221	218	208	211	218	224	223	219	215	224	244	159	0
0	0	0	0	18	44	82	107	189	228	220	222	217	226	200	205	211	230	224	234	176	188	250	248	233	238	215	0
0	57	187	208	224	221	224	208	204	214	208	209	200	159	245	193	206	223	255	255	221	234	221	211	220	232	246	0
3	202	228	224	221	211	211	214	205	205	205	220	240	80	150	255	229	221	188	154	191	210	204	209	222	228	225	0
98	233	198	210	222	229	229	234	249	220	194	215	217	241	65	73	106	117	168	219	221	215	217	223	223	224	229	29
75	204	212	204	193	205	211	225	216	185	197	206	198	213	240	195	227	245	239	223	218	212	209	222	220	221	230	67
48	203	183	194	213	197	185	190	194	192	202	214	219	221	220	236	225	216	199	206	186	181	177	172	181	205	206	115
0	122	219	193	179	171	183	196	204	210	213	207	211	210	200	196	194	191	195	191	198	192	176	156	167	177	210	92
0	0	74	189	212	191	175	172	175	181	185	188	189	188	193	198	204	209	210	210	211	188	188	194	192	216	170	0
2	0	0	0	66	200	222	237	239	242	246	243	244	221	220	193	191	179	182	182	181	176	166	168	99	58	0	0
0	0	0	0	0	0	0	40	61	44	72	41	35	0	0	0	0	0	0	0	0	0	0	0	0	0	0	0
0	0	0	0	0	0	0	0	0	0	0	0	0	0	0	0	0	0	0	0	0	0	0	0	0	0	0	0
0	0	0	0	0	0	0	0	0	0	0	0	0	0	0	0	0	0	0	0	0	0	0	0	0	0	0	0

In the following figure, we show a random sample of ten items from each kind of clothing. There is quite a variety in the images, as you can see, and not all of them are picture-perfect examples of each particular clothing class. That makes the problem somewhat more interesting. We want to create a neural network that takes as its input images like these and provides an output that tells us what kind of image it believes its input is.

Again, we'll build our neural network in layers. The first layer, comprising the input neurons, will have 784 neurons. Each one of them will take a single input, from a single patch in the image, and will simply output the value that it gets in its input. If the image is the ankle boot, the first neuron will get the value in the top-left patch, a 0, in its input, and it will output that 0. The rest of the neurons will get the values of the patches proceeding row wise, from top to bottom, left to right. The patch with the value 58, at the right end of the heel of the boot (the fourth row from the bottom, and the third column from the right) will get this 58 and copy it on its output. As rows and columns are counted in the neural network from the top and left, this neuron is in the twenty-fifth row from the top and twenty-sixth column from the left, making it the input neuron number $24 \times 28 + 26 = 698$.

The next layer will be densely connected to the input layer. It will consist of 128 ReLU neurons. This layer is not directly connected to the input images (the input layer is) and will not be directly connected to the output (we'll add another layer for that). Therefore it is a hidden layer, as we cannot observe it from the outside of the neural network. Being densely connected, this will result in a large number of connections between the input and hidden layer. Each neuron on the hidden layer will be connected to the outputs of all neurons on the input layer. There will be 784 input connections per neuron, for a total of $784 \times 128 = 100{,}352$ connections.

We will add another, last layer, which will contain the output neurons that will carry the results of the neural network. This will contain 10 neurons, one for each class. Each output neuron will be connected to all the neurons of the hidden layer, for a total of $10 \times 128 = 1{,}280$ connections. The grand total of all the connections between all the layers in the neural network will be $100{,}352 + 1280 = 101{,}632$. The resulting neural work will look, in schematic form, like the one on the next page. As it is impossible to fit all the nodes and edges, you can see dotted boxes standing for the bulk of nodes on the input and hidden layers; there are 780 nodes in the first box and 124 nodes in the second box. We have also collapsed the arrows going to the individual nodes inside the boxes.

The output of our neural network will consist of 10 outputs, one from each neuron on the layer. Each output neuron will represent one class, and its output will represent the probability that the input image belongs to this class; the sum of the probabilities of all 10 neurons will be 1, as it must happen when we deal with probabilities. This is an example of yet another activation function, called *softmax*, which takes as input a vector of real numbers and converts them to a probability distribution. Let's see the two examples that follow.

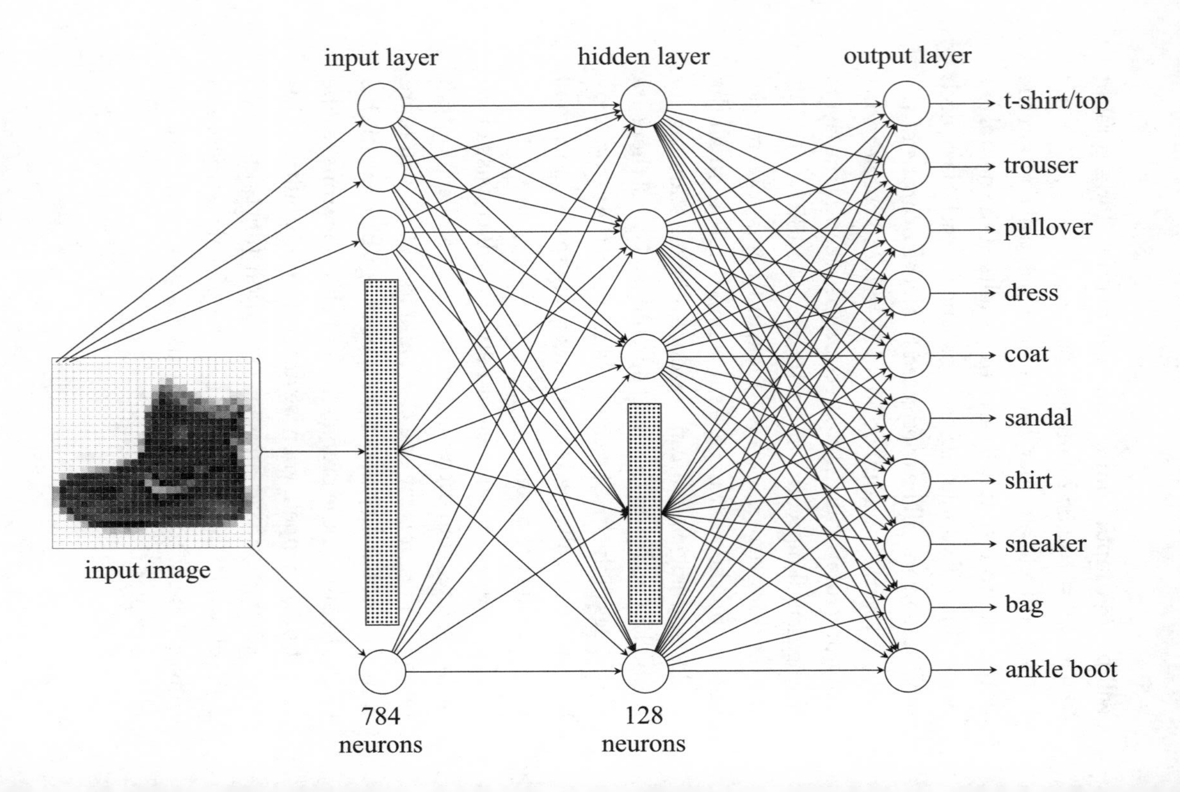
input layer
hidden layer
output layer
t-shirt/top
trouser
pullover
dress
coat
sandal
shirt
sneaker
bag
ankle boot
input image
784
neurons
128
neurons

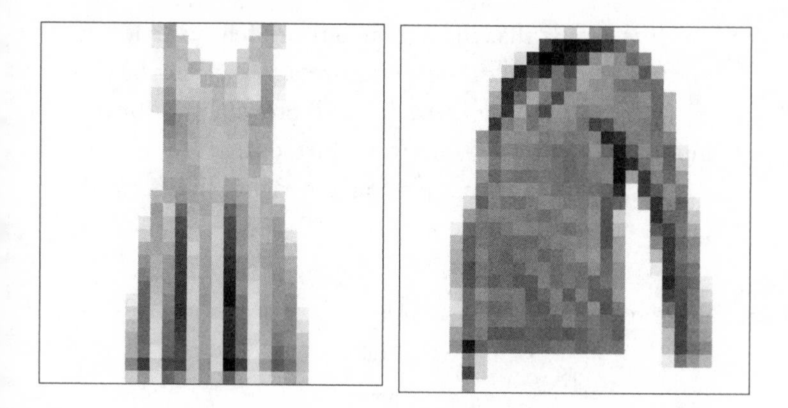

In the first example, on the left, after training we get this at the output of the network:

Output Neuron	Class	Probability
1	T-shirt/top	0.09
2	Trouser	0.03
3	Pullover	0.00
4	Dress	0.83
5	Coat	0.00
6	Sandal	0.00
7	Shirt	0.04
8	Sneaker	0.00
9	Bag	0.01
10	Ankle boot	0.00

That means that the neural network tells us that it is pretty certain it is dealing with a dress, giving it an 83 percent probability, leaving aside small probabilities for the input image being a T-shirt/top, shirt, or trouser.

In the second example, on the right, the network produces:

Output Neuron	**Class**	**Output**
1	T-shirt/top	0.00
2	Trouser	0.00
3	Pullover	0.33
4	Dress	0.00
5	Coat	0.24
6	Sandal	0.00
7	Shirt	0.43
8	Sneaker	0.00
9	Bag	0.00
10	Ankle boot	0.00

The neural network is 43 percent certain that it is dealing with a shirt—and it is wrong; the photo is really a picture of a pullover (in case you couldn't tell). Still, it did give its second best, at 33 percent, to the image being a pullover.

We gave one example where the network comes up with the right answer, and another instance where the network comes up with the wrong answer. Overall, if we give the network many images to recognize, all the 60,000 images in our training data set, we'll find out that it manages to get right about 86 percent of the 10,000 images in the test data set. That is not bad, considering that the neural network, even though it is way more complicated than the previous one, is still a simple one. From this baseline, we can create more complicated network structures that would give us better results.

Despite the increased complexity, our neural network learns in the same way as our simpler networks recognizing blobs of data and concentric circles. For each input during training we obtain an output, which we compare to the desired output to calculate the loss. The output now is not a single value but rather 10 values, yet the principle is the same. When the neural network recognizes a shirt with about 83 percent probability, we can compare that with the ideal, which would be to recognize it with 100 percent probability. Therefore we have two sets of output values: the one obtained by the network, with various probabilities assigned to the different kinds of clothes, and what we would like to have gotten from the network, which is a set of probabilities where all of them are zero apart from a single probability, corresponding to the right answer, which is equal to one. In the last example, the output contrasted to the target would be as follows:

Output Neuron	Class	Output	Target
1	T-shirt/top	0.00	0.00
2	Trouser	0.00	0.00
3	Pullover	0.33	1.00
4	Dress	0.00	0.00
5	Coat	0.24	0.00
6	Sandal	0.00	0.00
7	Shirt	0.43	0.00
8	Sneaker	0.00	0.00
9	Bag	0.00	0.00
10	Ankle boot	0.00	0.00

We take the last two columns and we calculate again a loss metric—only this time, as we do not have a single value, we do not calculate a simple squared difference. There exist metrics to calculate the difference between sets of values like these. In our neural network we used one such metric, called *categorical cross-entropy*, which indicates how much two probability distributions differ. Having calculated the loss, we update the neurons on the output layer. Having updated them, we update the neurons on the hidden layer. In short, we perform backpropagation.

We go through the same process for all images in our training data set—that is, for a whole epoch. When we are done, we do this all over again for another epoch. We repeat the process while trying to strike a balance: enough epochs so that the neural network will learn as much as possible from the training data set without going into too many epochs where the neural network will learn too much from the training data set. During learning, the network will be adjusting the weights and biases of its neurons, which are a lot. The input layer just copies values to the hidden layer, so no adjustments need to be done to the input neurons, but there are 100,352 weights on the hidden layer, 1,280 weights on the output layer, 128 biases on the hidden layer, and 10 biases on the output layer, for a total of 101,770 parameters.

Getting to Deep Learning

It can be proven that even though a neuron on its own cannot do much, a neural network can perform any computational task that can be described algorithmically and run on a computer. Therefore there is nothing that a computer can do that a neural network could not do. The whole idea, of course, is that we do not need to tell the neural network exactly how to perform a task. We only need to feed it with examples while using an algorithm to make the neural network learn how to perform the task. We saw that backpropagation is such an algorithm. Although we limited our examples to classification, neural networks can be applied to all sorts of different tasks. They can predict the values of a target quantity (for instance, credit scoring), translate between languages as well as understand and generate speech; and beat human champions in the game of Go, in the process baffling experts by demonstrating completely new strategies of playing a centuries-old game. They have even learned how to play the game of Go starting with just a knowledge of the rules, without access to a library of previously played games, and then proceeding to learn as if the neural network were playing games against itself.[8]

Today, successful applications of neural networks abound, yet the principles are not new. The Perceptron was invented in the 1950s, and the backpropagation algorithm is more than 30 years old. In this period, neural networks came and went out of fashion, with enthusiasm for their potential ebbing and flowing. What has really changed in the last few years is our capability to build really big neural networks. This has been achieved thanks to the advances in manufacturing specialized computer chips that can perform the calculations executed by neurons efficiently. If you picture all the neurons of a neural network arranged inside a computer's memory, then all the required computations can be carried out by operations on vast matrices of numbers. A neuron calculates the sums of the weighted products of its inputs; if you recall the discussion on PageRank in the previous chapter, the sum of the products is the essence of matrix multiplication.

It has turned out that *graphics processing units* (*GPUs*) are perfectly suited for this. GPUs are computer chips that are specially designed to create and manipulate images inside a computer; the term builds on *central processing units* (*CPUs*), the chip that carries out the instructions of a program inside a computer. GPUs are built to carry out instructions for computer graphics. The generation and processing of computer graphics requires numerical operations on big matrices; a computer-generated scene is a big matrix of numbers (think of the shoe). GPUs are the workhorses of game consoles. The same technology that arrests human intelligence in hours of diversion is also used to advance machine intelligence.

We started with the simplest possible neural network, consisting of a single neuron. Then we added a few neurons, and then we added a few more hundreds. Still, the image recognition neural network that we created is by no means a big one. Nor is its architecture complicated. We just added layer on layer of neurons. Researchers in the field of deep learning have made big strides in devising neural network designs. These architectures may comprise dozens of layers. The geometry of these layers need not be a simple one-dimensional set of neurons, like the ones we have here. For example, neurons inside a layer may be stacked on two-dimensional canvas-like structures. Moreover, it is not necessary to have each layer densely connected to the one before; other connection patterns are possible. Nor is it necessary to have the outputs of a layer simply connected to the inputs of the next layer. We may, for instance, have connections between non-consecutive layers. We may bundle up layers and treat them as modules, combining them with modules containing other layers to form more and more complex configurations. Today we have a menagerie of neural network architectures at our disposal, such that particular architectures are well suited for specific tasks.

The neurons on the layers in all the neural network architectures update the values of the weights and biases as they learn. If we reflect on what is happening, we can see that we have a set of inputs that transforms the layers during the learning process. Once the training stops, the layers have somehow, via the adjustments in their parameters, taken in the information represented by the input data. The weights and biases configuration of a layer represents the input it has received. The first hidden layer, which comes in direct contact with the input layer, encodes the neural network's input. The second hidden layer encodes the output of the first hidden layer, to which it is directly connected. As we proceed deeper and deeper into a multilayer network, each layer encodes the output received by the previous layer. Each representation builds on the previous one and therefore is on a higher level of abstraction from the one of the preceding layer. Deep neural networks, then, learn a hierarchy of concepts, proceeding to higher and higher levels of abstraction. It is in this sense that we talk of *deep* learning. We mean an architecture whereby successive levels represent deeper concepts, corresponding to higher levels of abstraction. In image recognition, the first layer of a multilayer network may learn to recognize small local patterns, such as edges in the image. Then the second layer may learn to recognize patterns that are built from the patterns recognized by the first layer, such as eyes, noses, and ears. The third layer may learn to recognize patterns that are built from the patterns recognized by the second layer, like faces. Now you can see that our neural network for recognizing the images was somewhat naive; we did not try to implement actual deep learning. By building abstractions on abstractions, we expect our network to find patterns that humans find, from structures in sentences, to signs of malignancy in medical images, to recognizing handwritten characters, to detecting online fraud.

Yet, you may say, it all boils down to updating simple values on simple building blocks—the artificial neurons. And you would be correct. When people realize that, sometimes they feel let down. They want to learn what machine and deep learning are, and the simplicity of the answer disappoints: something that appears to have human capabilities can be reduced to fundamentally elementary operations. Perhaps we would prefer to find something more involved, which would not fail to flatter our self-esteem.

We should not forget, however, that in science we believe that nature can be explained from first principles, and try to find such principles that are as simple as possible. That does not preclude complex structures and behaviors arising out of simple rules and building blocks. Artificial neurons are much simpler than biological ones, and even if the workings of biological neurons can be explained in simple models, it is thanks to the vast number of interconnected biological neurons that intelligence, as we know it, can arise.

This helps put some things into perspective. True, artificial neural networks can be uncanny in their potential. In order to make them work, however, an amazing amount of human creativity and terrific engineering effort is required. We have only scratched the surface in our account here. For instance, take backpropagation. That is the fundamental algorithm behind neural networks, allowing us to perform efficiently what is at heart a process of finding mathematical derivatives. Researchers have been busy devising efficient calculation techniques, such as *automatic differentiation*, a mechanism for calculating derivatives that has been widely adopted. Or take the exact way that changes in the neural network parameters are calculated. Various different *optimizers* have been developed, allowing us to deploy bigger and bigger networks that are at the same time more and more efficient. Turning to the underlying hardware, hardware engineers are designing better

Artificial neurons are much simpler than biological ones, and even if the workings of biological neurons can be explained . . . , it is thanks to the vast number of inter-connected biological neurons that intelligence . . . can arise.

and better chips to run more and more neural computations faster while using less computing power. Looking at network architectures, new neural network architectures are proposed that improve on existing ones. This is a hotbed of research and experimentation, and even encompasses efforts to build neural networks that design other neural networks. So every time you see a news report that a neural network has reached a new achievement, doff your hat to the hardworking people who made this possible.[9]

EPILOGUE

On July 15, 2019, Mark Carney, the Bank of England governor, presented the design of the new £50 note, expected to enter circulation about two years later. The Bank of England had decided in 2018 to celebrate a character from science with the new banknote and opened a six-week public nomination period for the selection. It received a total of 227,299 nominations for 989 eligible characters. From this, the Banknote Character Advisory Committee decided on a short list of 12 options. Then the governor made the final decision, selecting Alan Turing. He commented, "Alan Turing was an outstanding mathematician whose work has had an enormous impact on how we live today. As the father of computer science and artificial intelligence, as well as war hero, Alan Turing's contributions were far ranging and path breaking. Turing is a giant on whose shoulders so many now stand."[1]

Turing (1912–1954) was a genius who explored the limits and nature of computation, foresaw the rise of machines that would display intelligent behavior, grappled with the question of whether machines could think, contributed to mathematical biology and mechanisms of morphogenesis, and played a crucial role in the cryptanalysis of encrypted German messages during World War II (his contribution remained a secret for decades). In a tragic turn of events, Turing died by suicide. He had been arrested and convicted in 1952 for homosexuality, which was criminal in the United Kingdom at the time, and compelled to get hormonal treatment. An official pardon was issued in 2013. His appearance on the new note is a form of rehabilitation that would have been unthinkable a few decades back.[2]

Throughout this book we have been describing algorithms as consisting of simple steps, elementary enough that they can be carried out using a pen and paper. Given that we implement algorithms in computer programs, the question of what really is an algorithm will help us understand what can really be computed. This requires us to dig deeper into the nature of these simple steps. After all, what a primary school student can do with a pen and paper is different than what a college graduate can do. Is it possible to define precisely what kind of steps an algorithm could be made of? Turing offered an answer even before digital computers were built. He proposed a model machine in 1936 in order to answer the question of what a computer, any computer, can do. A *Turing machine* is a simple contraption. It consists of the following parts:[3]

1. A *tape*. The tape is divided into squares or *cells*. Each cell can be blank or contain a symbol from some alphabet. The tape can be infinitely long.

Is it possible to define precisely what kind of steps an algorithm could be made of? . . . [Turing] proposed a model machine in 1936 in order to answer the question of what a computer, any computer, can do.

2. A *head* that can move left and right along the tape, one position at a time. The head can read the symbol in the cell underneath. We call the symbol in that cell the *scanned symbol*. The head can erase or overwrite the scanned symbol.

3. A *finite control*, also called a *state register*. The finite control can be in any of a finite set of states. You can think of it as a dial inscribed with states, and an indicator that can point to any one of them.

4. A *finite instructions table*. Each instruction specifies the next *move* of the machine. This is what the machine will do, given its current state and the scanned symbol.

You can see a Turing machine in the figure on the next page.[4]

The alphabet of this particular Turing machine consists of 1 and $\star$. The finite control shows that the machine can be in one of seven states, $q_0, q_1, \ldots, q_6$. The instructions table has one row for each possible state, and one column for each possible symbol; we use *B* to stand in for blank so that we can see it. The current state is indicated by the row, and the scanned symbol by the column. Each entry in the instructions table contains a triplet, describing a move, or a dash, meaning that the machine has nothing to do in this row and column combination.

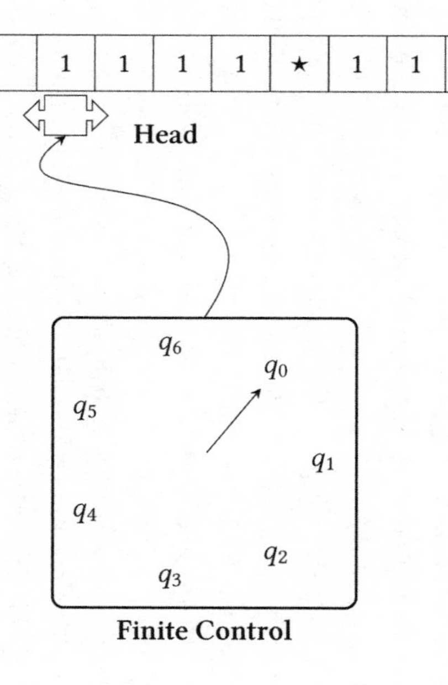

Finite Instructions Table

State	Symbol: 1	Symbol: ★	Symbol: B
q_0	(q_1, B, R)	(q_5, B, R)	–
q_1	$(q_1, 1, R)$	$(q_2, \star, R)$	–
q_2	$(q_3, \star, L)$	$(q_2, \star, R)$	(q_4, B, L)
q_3	$(q_3, 1, L)$	$(q_3, \star, L)$	(q_0, B, R)
q_4	$(q_4, 1, L)$	(q_4, B, L)	$(q_6, 1, R)$
q_5	(q_5, B, R)	(q_5, B, R)	(q_6, B, R)
q_6	–	–	–

A move of the machine consists of three actions:

1. The machine may change or remain in the same state. The new state is the first element of the triplets in the finite instructions table.

2. It will write a symbol under the head. The symbol may be the same with the one already there (then the result is that the existing symbol remains in the cell). The symbol to be written is the second element of the triplets.

3. The head will shift either to the left (*L*) or right (*R*) of the current cell. The shift is the third element of the triplets.

Our example Turing machine executes an algorithm that computes the difference of two numbers a and b when $a > b$; otherwise, it returns zero. This operation is called *monus* or *proper subtraction*, and we write $a \dot{-} b$. We have $4 \dot{-} 2 = 2$ and $2 \dot{-} 4 = 0$.

Initially, we place the machine's *input* on the tape. The input is a finite string of symbols from the machine's alphabet. All other cells of the tape, to the left and right of it, are blank. In this Turing machine, the input is $1111 \star 11$. The input represents the numbers four and two in the *unary numeral system*, separated by $\star$.

This machine starts with its head on the leftmost input cell. The finite control points at the q_0 state. Then the machine starts working and performs its moves. If we follow the machine's operation for the first six moves, we'll see that it goes like this:

1. We are at state q_0 and the scanned symbol is 1:

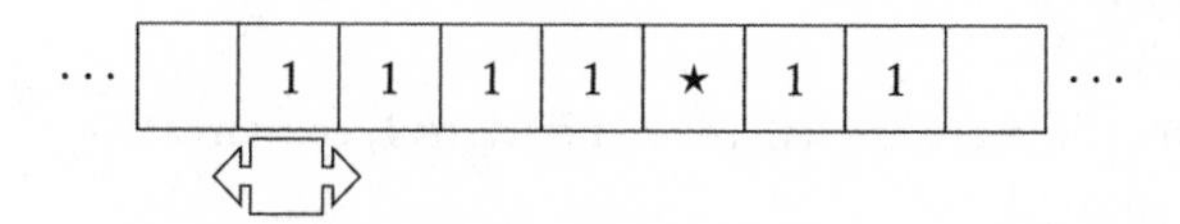

   The instruction table gives us (q_1, B, R), so the machine will change its state to q_1, overwrite 1 with blank, and move right. The tape and head will be:

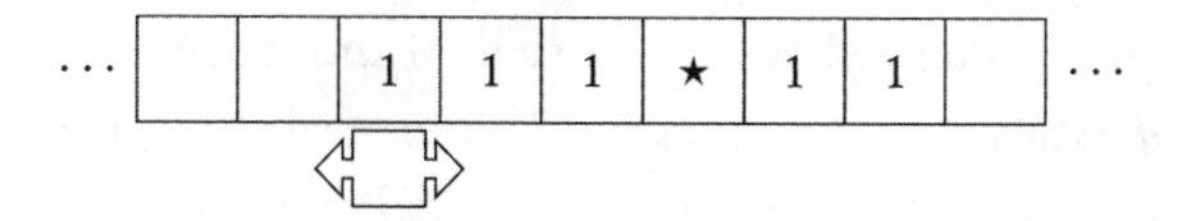

2. For the q_1 state and scanned symbol 1, the instruction table gives us $(q_1, 1, R)$. The machine will read and write 1, leaving the cell as it is, and will move right, remaining at state q_1:

3. The machine does the same as step 2, reading and writing 1, remaining at q_1, and moving right:

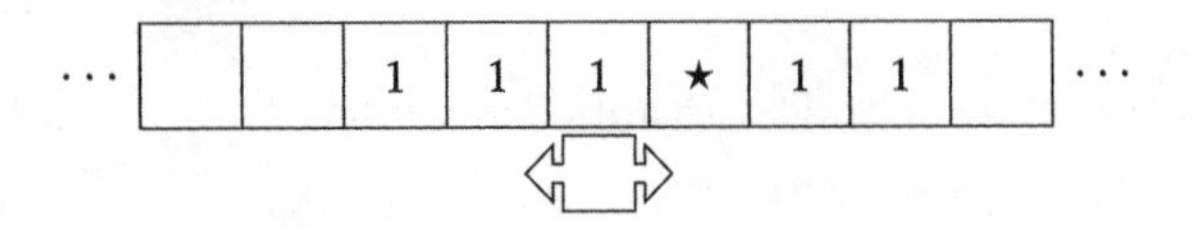

4. Again, the machine will read and write 1, remain at q_1, and move right:

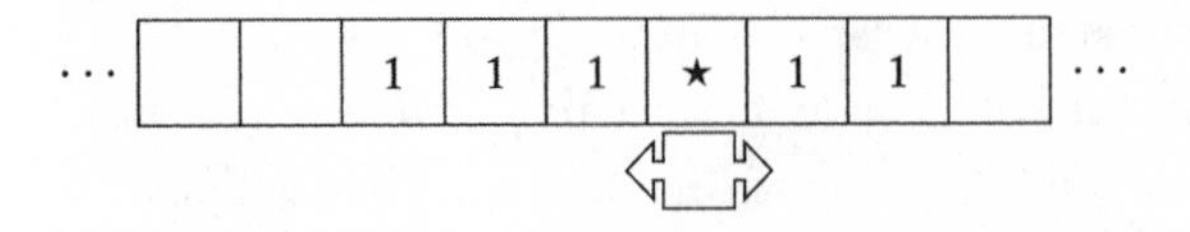

5. The head has moved over the $\star$ symbol and remained at state q_1. The instruction is $(q_2,\star,R)$. The machine will change state, to q_2, leave $\star$ on the tape, and move right:

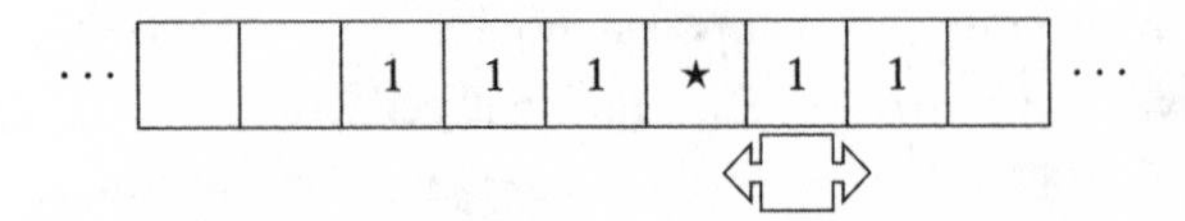

6. The head has moved over the 1 to the right of $\star$ and is at state q_2. The instruction is $(q_3,\star,L)$. The machine will change state, to q_3, write $\star$ over the 1, and move back left:

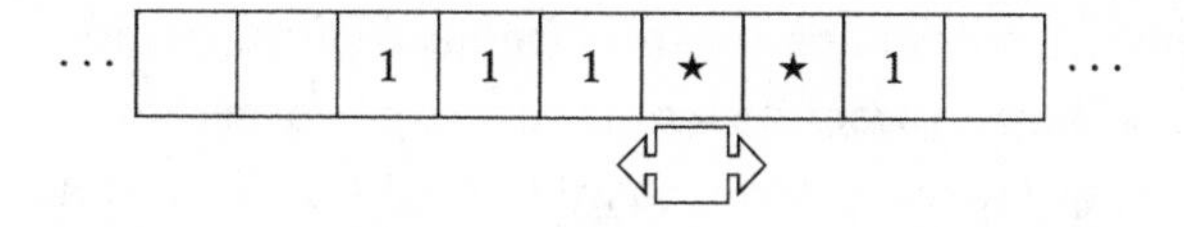

The machine will continue working in this way, performing the moves prescribed by the instruction table. If we take a higher-level view, we'll realize that the machine executes a loop. In each iteration, it finds the leftmost 1 and replaces it with a blank. It then searches right for a $\star$. When it finds it, it continues going right until it finds a 1, which it turns into a $\star$. Therefore in each iteration, the machine strikes out a 1 on the left and right of $\star$. At some point, this will no longer be possible. Then the machine will replace all $\star$ symbols with blanks and will terminate. The tape will contain 11, equivalent to the number 2, surrounded by blanks. To indicate termination, the machine enters the state q_6, where according to the instructions table there is nothing to do, and it stops.

If we provide as input $11 \star 1111$, the machine will beaver away until it stops with a tape full of blanks, which is equivalent to 0. If we give the machine any input consisting of a ones followed by an asterisk and then b ones, it will follow its moves until it leaves the tape with either $a - b$ ones, if $a > b$, or otherwise all blanks.

This Turing machine executes an algorithm for computing the monus operation based on its input and following the instructions described in its instructions table.

The steps are so elementary that the head of the Turing machine scampers around a lot in order to perform the operation. It will take 21 moves to find that $2 \dot{-} 4 = 0$ and 34 moves to find that $4 \dot{-} 2 = 2$. But how simple these moves are! Anybody with a modicum of intelligence can carry them out. The rudimentary nature of the steps is exactly the point. You do not need any advanced qualifications to perform the steps of a Turing machine; you only need to look up a table, move around on a tape, read and write one symbol at a time, and keep track of what your state is. That is all. Yet it is not trivial because the answer to the question of what kind of steps an algorithm could be made of, is that they are the steps that a Turing machine could perform.

In this book we have been describing algorithms at a higher level, with more complex steps. That is for our convenience because a Turing machine works at such a low level of detail that it would be unwieldy to use it to describe our algorithms. But all the steps of all the algorithms we have depicted could be presented as steps of a properly constructed Turing machine. We have described a simple Turing machine to implement the monus operation. For a more complex algorithm we would need a Turing machine with more states, a bigger alphabet, and a bigger instructions table. But we could still build it, if we wanted.

The simplicity of the Turing machine belies its ambit; given any algorithm, we can construct a Turing machine that implements it. As computers run algorithms, any algorithm that is computable by a computer is computable by a Turing machine. Or in other words, *whatever we can do with an algorithm, we can do with a Turing machine*. That is a loose rendering of the *Church-Turing thesis*, named after Turing and the US mathematician Alonzo Church (1903–1995), one of the founders of theoretical computer science. It being a *thesis*, it is not something that has been proved, and we do not know if it can be proved mathematically. It is theoretically possible that it could be disproved, if somebody devises some alternative form of computation that computes things that a Turing machine cannot compute. We do not believe this is likely to happen. We therefore take the Turing machine to be a formal description of the notion of an algorithm.[5]

You can imagine any computer, as powerful as you want it. The computer will be way faster than a Turing machine that operates on a tape of symbols as we have described it. But everything it calculates algorithmically, a Turing machine can calculate too. You can even imagine computers that we have not been able to manufacture yet. Our computers work with *bits*, which can exist in only two states, 0 and 1. *Quantum computers* work with *qubits*. When we examine the state of a qubit, this will be 0 or 1, like a bit. Yet a qubit, when we don't examine it, can be in a combination, called *superposition*, of the two binary states 0 and 1. It is as if a qubit is both 0 and 1, until we decide to read it, when it decides to be one of these two values. This allows quantum computers to represent multiple states of computation at once. A quantum computer would allow us to solve fast problems that are not easily solved by classical computers. Unfortunately, building a quantum computer is difficult with the current technology. And even a quantum computer could not do something that a Turing machine cannot do. Even though it would be able to solve some problems more efficiently than any existing classical computer, or any Turing machine for that matter, it still won't be able to solve any problems that a Turing machine cannot solve.

Our computational limits are given by Turing machines. Anything a computer can do, we could really do with a pen and paper, working on a tape of symbols. Everything you see executed on any digital device is, in essence, a series of such elementary symbol manipulations. In the natural sciences, we behold the world and believe that we can explain it using fundamental principles. In computing, it is the other way around. We have our fundamental principles and believe that we can do amazing feats with them.

When Turing proposed his machine as a model for computation, digital computers did not even exist. That did not prevent him from exploring the capabilities of computing machines that would be created in the future. When we think about the limits of computers, we should

Our computational limits are given by Turing machines. Anything a computer can do, we could really do with a pen and paper. . . . Everything you see executed on any digital device is . . . a series of such elementary symbol manipulations.

also keep in mind that inside these limits, the human intellect has created wonders. The limits of computation have not curtailed our creativity to continue developing algorithms for every aspect of our lives. When writing was invented in Mesopotamia, its purpose was to aid record keeping, not write literature. The first writers were probably accountants, not authors, yet from such humble beginnings emerged William Shakespeare. Who knows what, in time, algorithms will bring.

GLOSSARY

activation (neuron)
The emission of output from a neuron.

activation function
A function that determines the output of a neuron based on its input.

acyclic graph
A graph that has no cycle.

adjacency matrix
A matrix that represents a graph. It has a row and column for each vertex of the graph. Its contents are 1 in each entry whose row and column correspond to two vertices connected by an edge in the graph; all other entries are 0.

algorithm

1. Go to the first page of the book.
2. Read the current page.
3. If you don't understand, go to step 2. Otherwise go to step 4.
4. If there is a next page, make it your current page and go to step 2. Otherwise terminate.

approximation
Solving a problem by using an algorithm that may not find the optimal solution, but one that is not far from it.

automatic differentiation
A set of techniques to evaluate the derivative of a function numerically—that is, not analytically, which would entail using the calculus rules for differentiating functions.

backlink
A link that points to the web page we are visiting, and by extension, the web pages that contain links that point to the web page we are visiting.

backpropagation algorithm
A fundamental algorithm for training neural networks. The network corrects its configuration (its weights and biases) by propagating adjustments from the final layer back toward the first layer.

bias
A numerical value attached to a neuron that controls its propensity to fire.

big O
A notation for computational complexity. Given an algorithm and input greater than some threshold, it gives us an upper bound on the expected number of steps required by the algorithm to complete. We want the input to be larger than some threshold because we are interested in the behavior of an algorithm on large data. The big O complexity for an algorithm gives us a guarantee that for large data, the algorithm will not require more than a particular number of steps. For example, a complexity of $O(n^2)$ means that for input of size n that exceeds some threshold, the algorithm will not take more than a constant multiple of n^2 steps to complete.

binary search
A search algorithm that works on ordered data. We check the item in the middle of the search space. If it matches the one we are looking for, we are fine. Otherwise, we repeat the procedure to the left or right half, depending on whether we have overshot or undershot our target.

bit
The basic unit of information stored on a computer. A bit can take one of two values, 0 or 1. The word bit comes from binary digit.

bug
An error in a program. The term bug was used by Thomas Edison for a technical fault. In the early days of computing, real bugs would make their way into the machinery, causing them to fail. A moth that did that was found inside the Harvard Mark II computer in 1947. The moth has been preserved in the machine's logbook, which is part of the collection of the Smithsonian National Museum of American History.

categorical cross-entropy
A loss function that calculates the difference between two probability distributions.

central processing unit (CPU)
The chip that carries out the instructions of a program inside a computer.

chromatic index
In graph coloring, the minimum number of colors required to color the edges of the graph.

Church-Turing thesis
The hypothesis that everything that can be computed by an algorithm can be computed by a Turing machine.

classifier
A program that classifies an observation in one out of a number of possible classes.

complexity (computational complexity)
The time required for an algorithm to run. The time is expressed on the order of elementary computational steps required to complete.

complexity class
A set of problems that require the same amount of a resource (such as time or memory) to be solved.

control structure
The three ways in which steps can be combined in an algorithm or program: sequence, selection, and iteration.

cycle
In graphs, a path that starts and end at the same node.

dangling node
In the PageRank algorithm, a node with only incoming edges and no outgoing edges.

data structure
A way to organize data, such that we can handle the data with a set of specific, prescribed operations.

decision boundary
The values of one or more variables that form the boundary between two different outcomes of a single decision based on the variable or variables.

deep learning
Neural networks that consist of many hidden layers, arranged such that succeeding layers represent deeper concepts, corresponding to higher abstraction levels.

degree (node)
The number of edges adjacent to a node.

densely connected
Layers in a neural network arranged such that all the neurons of a layer are connected to all the neurons of the following layer.

derivative
The slope of a function at a point; equivalently, the rate of change of a function. For example, acceleration is the derivative of speed (the rate of change of speed in time).

Dijkstra's algorithm
An algorithm invented in 1956 by a young Dutch computer scientist, Edsger Dijkstra, to find the shortest path between two nodes in a graph. It works with graphs that contain positive weights.

directed graph
A graph in which the edges are directed. A directed graph is also called a digraph for short.

divide and conquer
A problem-solving method where we solve a problem by breaking it into smaller problems (typically two) and then do the same on the smaller problems, until the problems get so small that the solution is straightforward to find.

edge coloring
The assignment of colors to the edges of a graph so that no two adjacent edges share the same color.

eigenvector

In linear algebra, an eigenvector is a vector that, when we multiply it by a specific matrix, the result is the same vector multiplied by a number; that number is its eigenvalue. PageRank finds the first eigenvector of the Google matrix—that is, the eigenvector of the Google matrix with the largest eigenvalue, which is equal to one.

epoch

In machine learning, a pass, during training, through the whole training data set.

Euclid's algorithm

An algorithm for finding the greatest common divisor of two integers, presented in the *Elements*, a set of 13 books written by the ancient Greek mathematician Euclid (ca. 300 BCE). The *Elements* treats geometry and number theory, starting from axioms and proving theorems based on the axioms. It is the oldest extant work of mathematics that uses this deductive approach, and as such, one of the most influential books in the history of science.

Eulerian path

A trail through a graph such that each edge is visited exactly once. It is also called a Euleurian walk.

Eulerian tour

A Eulerian path that starts and ends at the same node. It is also called a Eulerian tour.

Euler's number

The mathematical constant e, approximately equal to 2.71828. It is the limit of $(1 + 1 / n)^n$ as n approaches infinity.

execution path

The series of steps that an algorithm carries out during its execution.

exponential growth

A growth pattern in which a number of things is successively multiplied by itself. For example, we may start with a things, and then we'll get $a \times a$ things, then $a \times a \times a$, and in general $\overbrace{a \times a \times \cdots \times a}^{n} = a^n$. Numbers grow fast with exponential growth.

factorial
The factorial of a natural number n is the product of all numbers from 1 up to and including n. We use the symbol $n!$ so we have $n! = 1 \times 2 \times \cdots \times n$. The definition can be extended to all real numbers, but that does not concern us here.

factorial complexity
Computational complexity that follows factorial growth. In big O notation, it is $O(n!)$.

fire (neuron)
See activation (neuron).

fitting
In machine learning, the process of learning from the data. In this process we construct a model that fits the observations.

garbage in, garbage out
If we feed a program garbage, instead of its expected input, we should expect no miracles: the program will produce garbage instead of its expected output.

global optimum
The best overal solution to a problem.

Google matrix
A special kind of matrix (a modification of the hyperlink matrix) that is used in the power method in the PageRank algorithm.

gradient
A vector containing all the partial derivatives of a function.

graph
A set of nodes, also called vertices, and edges, also called links, connecting them. Graphs can be used to model any kind of linked structure, from people to computer networks. As a result, many problems can be modeled as graphs, and many algorithms have been developed that work on top of them.

graph coloring
The edge or vertex coloring of a graph.

graphics processing unit (GPU)
A chip specially designed to handle the instructions for the creation and manipulation of images inside a computer.

greatest common divisor (gcd)
Given two integers, the largest integer that divides both.

greedy algorithm
An algorithm in which when we have to choose between alternative courses of action, we choose the one that gives us the greatest immediate payoff. This does not necessarily lead to the optimum outcome in the end.

hardware
The physical components that make up a computer or digital device. The term complements software.

head
The first item in a list.

heuristic
A strategy for making choices among alternatives in an algorithm. A greedy heuristic would require us to take the option that looks best right now (never mind what could happen in the future).

hidden layer
A neural network layer that is not directly connected to the input or output of the network.

Hierholzer algorithm
An algorithm for finding Eulerian circuits on graphs. It was published by the German mathematician Carl Hierholzer in 1873.

hill climbing
A metaphor for describing problem solving. The solution is at the top of the hill, and we have to climb from its foot. At each step there may be a decision to take among alternative paths. Depending on our choices, we may select the best path overall, a path that is not the best but still takes us to the top, or alas a path that leads to a plateau. If the worst happens and we reach a plateau, we'll have to go back to a previous position to start moving along a different path.

hyperlink
A reference from a text to another part of the text or a different text. On the web, hyperlinks are links between web pages that the user may follow while browsing.

hyperlink matrix
A matrix representing the structure of a graph; it is like an adjacency matrix, but we divide the elements of its row by the number of nonzero elements in the row.

hyperplane
The generalization of the plane in more than three dimensions.

hypertext
Text that contains hyperlinks.

image recognition
The computational task of recognizing patterns in images.

insertion sort
A sorting method where we take each item and insert it into its correct position among the already sorted items.

internet
A global network of computers and digital devices, interconnected by means of a common suite of communication protocols. Initially, it was with its first letter capitalized (Internet) because internet could refer to any network that extended beyond the internal confines of an institution, which is called an intranet. As the global internet took off, however, the initial capital fell out of favor, probably saving a significant amount of ink.

intractable problem
A problem for which the best algorithms we know will take an inordinate amount of time to handle anything but trivial cases.

iteration
See loop.

key

A part of a record that we use for sorting or finding it. A key may be atomic, when it cannot be decomposed into smaller parts (for instance, an identification number), or composite, when it consists of smaller pieces of data (like the full name comprising first name, middle name, and surname).

label

In machine learning, a value representing the category to which an observation belongs. In training, the computer is given problems along with their solutions; when the problem is classification, the solutions are the labels representing the classes.

linear search

A search algorithm in which we examine each item in turn until we find the one we are looking for. It is also called a sequential search.

linear time

Time proportional to the input of an algorithm, written as $O(n)$.

linearly separable

A data set whose observations can be separated into two categories by a straight line in two dimensions, plane in three dimensions, or hyperplane in more dimensions.

list

A data structure that contains items. Each item points to the next one, apart from the last item, which points nowhere, or to null, as we say. The items are therefore linked to each other, and such a list is also called a linked list.

local optimum

A solution that is better than all the other neighboring solutions, but not the overall best. A neighboring solution is a solution in which we can get with a single move from the solution we are now.

logarithm

The inverse of raising to a power. The logarithm is the answer to the question, "To which power should I raise a number to get the value I want?" If we ask, "To which power should I raise 10 to get 1,000?," the answer is 3 because $10^3 = 1{,}000$. The number we will raise to the power is called the base of the logarithm. We write $log_a x = b$ if $a^x = b$. For $a = 2$ we write lgx.

logarithmic time
Time proportional to the logarithm of the input of an algorithm—for example, *O*(*lgn*). Good searching algorithms take logarithmic time.

loglinear time
Time proportional to the product of the size of the input and logarithm of the input of an algorithm—for example, *O*(*nlgn*). Good sorting algorithms take loglinear time.

loop
A sequence of instructions in a computer program that is repeated. A loop ends when a condition is fulfilled. A loop that does not end is an infinite loop and is usually a bug because it may lead to a program that fails to terminate. *See* iteration.

loss
The difference between the actual and desired output of a machine learning algorithm. It is typically calculated by a loss function.

machine learning
The use of algorithms that solve problems by learning automatically from examples.

matrix
A rectangular array, typically of numbers or more generally mathematical expressions. The contents of a matrix are arranged horizontally in rows and vertically in columns.

Matthew effect
The phenomenon of the rich getting richer and poorer getting poorer. Named after the Gospel of Matthew (25:29), it has been found to apply to many contexts, not just material wealth.

minimization problem
A problem in which, among the possible solutions, we try to find the one with the minimum value.

merge sort
A sorting method that works by repeatedly merging larger and larger sets of sorted items.

Moore's law
The observation, made in 1965 by Gordon Moore, founder of Fairchild Semiconductor and Intel, that the number of transistors in an integrated circuit doubles about every two years. It is an example of exponential growth.

move to front
A self-organizing search algorithm. When we find the item we are looking for, we move it to the first position.

multigraph
A graph in which an edge can occur more than once.

multiset
A set in which an element can appear multiple times; in mathematics, in a normal set an element cannot appear more than once.

node
An item in various data structures. Items in lists are called nodes.

neuron
A neuron is a cell that forms the basic building block of the nervous system. It receives signals from other neurons and propagates them to other neurons in the nervous system.

null
Nothingness in a computer.

online algorithm
An algorithm that does not require the full input to a problem in order to produce a solution. An online algorithm gets the input incrementally, as this arrives, and at each point produces a solution that takes account of the input it has received so far.

onset
The accented part of a rhythm.

optimal stopping problem
The problem of knowing the best time to stop when you are trying to maximize a reward or minimize a penalty.

optimizers
Algorithms that optimize the value of a function. In machine learning, optimizers typically minimize the value of the loss function.

overfitting
The equivalent of learning by rote in machine learning. The model that we are trying to train follows the training data so closely that it fits them too well. As a result, it does not predict correct values for other, unknown data.

overflow
Going beyond the range of allowable values on a computer.

PageRank
An algorithm used to rank web pages in terms of their importance. It was developed by the founders of Google and was the foundation of the Google search engine. The rank of a web page is its pagerank.

pagerank vector
A vector containing the pageranks of a graph.

partial derivative
In a function of many variables, the derivative of the function with respect to one variable, holding all other variables constant.

path
In a graph, a sequence of edges that connect a sequence of nodes.

path length
The sum of the weights along a path in a graph. If a graph does not have weights, it is the number of the links constituting the path.

Perceptron
An artificial neuron that uses the step function for its activation.

permutation
A rearrangement of some data in a different order.

pointer
A place in computer memory that holds the address of another place in computer memory. In this way, the former points to the latter.

polynomial time
Time proportional to the input to an algorithm raised to a constant power, such as $O(n^2)$.

power method
An algorithm that starts with a vector, multiplies it by a matrix, and then repeatedly multiplies the result by the matrix until it converges into a stable value. The power method is at the heart of PageRank; the vector at which it converges is the first eigenvector of the Google matrix.

program
A set of instructions, written in a programming language, that describes a computational process.

programming
The art of writing computer programs.

programming language
An artificial language that can be used to describe computational steps. A programming language can be executed on a computer. Like a human language, a programming language has syntax and grammar, specifying what can be written in it. Several programming languages exist, and new programming languages are developed all the time in an effort to make programming more productive (or because many people cannot resist creating their own language and hope it will be widely adopted). A programming language can be high level, when it looks somewhat akin to a human language, or low level, when its constructs are rudimentary, mirroring the underlying hardware.

punched card
A piece of stiff paper that records information by the location of the punched holes on it. It is also called a punch card. The cards were used in early computers, and before that, in machines such as Jacquard looms, in which they described the pattern to be woven.

quantum computer
A computer that leverages quantum phenomena to perform computations. Quantum computers work with qubits instead of bits. Some problems can be solved much faster on quantum computers than on classical ones. The manufacture of quantum computers presents difficult physical challenges.

qubit
The basic unit of quantum information. A qubit can exist in a superposition of two states, 0 and 1, until we measure it, when it collapses to one of the two binary values. A qubit can be implemented using quantum properties, such as the spin of an electron.

quicksort
A sorting method that works by repeatedly selecting an item and moving the other items around it so that all smaller items are on the one side and all the rest on its other side.

radix sort
A sorting method that works by breaking the keys into their parts (for example, digits for numerical keys) and placing the items into piles corresponding to the values of their parts (ten piles, one for each digit). We start by forming piles based on the last digit, then we stack all piles and redistribute to piles based on the one but last digit, and so on. When we do the procedure for the first digit, we end up with a sorted pile. It is a string sorting method because we treat numerical keys as a string of digits.

random surfer
A person who surfs the web by going from page to page, choosing the next page according to the probability given by the Google matrix.

randomization
The use of randomness in algorithms. In this way, an algorithm may be able to find good solutions to a problem in most cases, even if it would be computationally infeasible to find the optimal solution.

record
A set of related data describing an entity for a particular application. For example, a student record can include identification data, enrollment year, and transcripts.

rectifier
An activation function that turns all negative inputs to zero, or otherwise its output is directly proportional to its input.

relaxation
A method in graph algorithms, where we assign the worst possible value to the values we want to find, and the algorithm proceeds by producing better and better estimates for these values. We therefore start with the most extreme values possible, and gradually relax them with values that are closer and closer to the final result.

ReLU
A neuron that uses a rectifier as its activation function. ReLU stands for rectified linear unit.

search space
The domain of values in which we search.

secretary problem
An optimal stopping problem. From a pool of candidates, we examine each one in turn. We must make the decision to hire or not on the spot, without being able to reverse past decisions, and without having examined the remaining candidates.

selection
In algorithms and programming, a choice, based on some logical condition, between alternative series of steps to be executed.

selection sort
A sorting method where each time we find the minimum of the unsorted items and put it into its correct position.

self-organizing search
Search algorithms that take advantage of the popularity of search items by moving them to positions where we'll be able to find them faster.

sequence
In algorithms and programming, a series of steps executed one after the other.

shortest path
The shortest path between two nodes in graph.

sigmoid
An S-shaped function whose values range from 0 to 1.

social network
A graph in which nodes are people, and the edges are the relationships between them.

softmax
An activation function that takes as input a vector of real numbers and turns it into another vector that is a probability distribution.

software
The set of programs running on a computer or digital device; the term complements hardware. The terms have been used before computers in a different setting. In 1850, rubbish-tip pickers were using the terms "soft-ware" and "hard-ware" to distinguish between material that would decompose and everything else. These meanings may bring solace to anybody struggling with a computer that won't do what it is supposed to do.

spallation
Breaking a material into smaller pieces. In nuclear physics, the material is a heavy nucleus that emits a large number of protons and neutrons after being bombarded with a high-energy particle.

sparse matrix
A matrix in which most elements are equal to zero.

string
A sequence of symbols. Traditionally a string was a sequence of characters, but nowadays what can go into a string depends on the actual application; it may be digits, alphabetic characters, punctuation, or even more recently invented symbols such as emojis.

string sorting method
A sorting method that treats its keys as a sequence of symbols. For example, the key 1234 is treated as the string of symbols 1, 2, 3, 4 instead of the number 1,234.

supervised learning
A machine learning approach in which we provide an algorithm with input problems accompanied by their solutions.

synapse
A connection between neurons.

tabulating machine
Electromechanical devices that could read punched cards and use the information on them to produce a tally.

tanh (hyperbolic tangent)
An activation function that looks like the sigmoid function, but its output ranges from −1 to 1.

test data set
Data that we set aside during training so that we can use them to check how well a particular machine learning approach will perform with real-world data.

tour
A path that starts and ends at the same node in a graph. It is also called a circuit.

training
In machine learning, the process of providing an algorithm with example inputs so that it can learn to produce correct outputs.

training data set
Data that we use with machine learning algorithms to train them to solve problems.

transposition method
A self-organizing search algorithm. When we find an element, we swap it with the one preceding it. In this way, popular items are moving to the front.

traveling salesman problem
Also known as the traveling salesperson problem, but people did not put much thought into gender definitions. The problem that asks us, If we have a list of cities and the distances between each pair of them, what is the shortest possible route that one should take to visit each city once and return to the origin city? It is probably the most famous intractable problem.

Turing machine
An idealized (abstact) machine, described by Alan Turing, consisting of an infinite tape and movable head that reads and writes symbols on the tape following a set of prescribed rules. The Turing machine can implement any algorithm and therefore can be used as a model of what can be computed.

unary numeral system
The number system using a single symbol for representing numbers; for instance, a stroke representing a unit, so that III represents three.

undirected graph
A graph in which the edges are undirected.

unsupervised learning
A machine learning approach in which we provide an algorithm input problems without their solutions. The machine learning algorithm then must derive what the expected input should be in order to be able to produce it.

vector
A horizontal row or vertical column of numbers (or more generally, mathematical expressions). Usually we meet vectors in geometry, where it is a geometric entity with a length and direction, represented as a row or column containing their numerical coordinates; however, the notion of a vector is more general than that—take, for example, the pagerank vector. A vector is a special case of a matrix.

vertex coloring
The assignment of colors to the vertices of a graph so that no two adjacent vertices share the same color.

weight (graph)
A number attached to an edge of a graph. Such a number may, for example, model a reward or penalty associated with the link between the nodes connected by the edge.

weight (neuron)
A numerical value attached to a synapse in a neuron. From each synapse, the neuron receives an input multiplied by the weight of the synapse.

weighted input (neuron)
The sum of the products of the inputs with the weights of a neuron.

NOTES

Preface

1. For these and more indicators of the global progress achieved through the ideas of the Enlightenment, see Pinker 2018.

Chapter 1

1. "The Algorithmic Age" was aired on February 8, 2018, on *Radio Open Source*.
2. For an account of algorithms in ancient Babylon, see Knuth 1972.
3. The algorithm for distributing a number of pulses in timing slots in the SNS was given by Eric Bjorklund (1999). Godfried Toussaint (2005) noticed the parallel with rhythms, and his work is the basis for our exposition. For a more extensive discussion, see Demaine et al. 2009. For a book-length treatment of algorithms and music, see Toussaint 2013.
4. The criteria come from Donald Knuth (1997, sec. 1), who also starts his exposition with Euclid's algorithm.
5. For a discussion of the enumeration of the paths on the grid, see Knuth 2011, 253–255; it is the source for the example and path images. For the algorithm that gives the number of possible paths, see Iwashita et al. 2013.
6. For these number descriptions, see Tyson, Strauss, and Gott 2016, 18–20. In Dave Eggers's novel *The Circle*, a thinly disguised technology company calculates the number of grains of sand in the Sahara Desert.
7. To fold paper n times, the paper must be large enough. If you fold it always along the same dimension, you will need a long sheet of paper. The length is given by the formula $L = \frac{\pi t}{6}(2^n + 4)(2^n - 1)$, where t is the paper's thickness and n is the number of folds. If you fold a square sheet of paper in alternate directions, then the width of the square must be $W \approx \pi t 2^{(3/2)(n-1)}$. The reason why the formulas are more complicated than simple powers of two is that every time you fold the paper, you lose some part of it as it curves along the edge of the fold; it's from calculating these curves that π enters the picture in these formulas. The formulas were found in 2002 by Britney Crystal Gallivan, then a junior in high school. She went on to demonstrate that a 1,200 meters–long sheet of toilet paper could be folded in half 12 times. For a nice introduction to the power of powers (including this example), see Strogatz 2012, chapter 11.

8. "Transistor Count," Wikipedia, https://en.wikipedia.org/wiki/Transistor_count.

9. That is because to compare n items between them, you need to take one of them and compare it to all the other $n - 1$ items, then you take another one and compare it to the other $n - 2$ items (you have already compared it to the first item you used), and so on. That gives $1 + 2 + \cdots + (n - 1) = n(n - 1) / 2$ comparisons. Then you get $O(n(n - 1) / 2) = O(n^2 - n / 2) = O(n^2)$, because according to the definition of big O, if your algorithm runs in time $O(n^2)$, it will certainly run in time $O(n^2 - n / 2)$.

Chapter 2

1. Image retrieved from the Wikipedia Commons at https://commons.wikimedia.org/wiki/File:Konigsberg_Bridge.png. The image is in the public domain.

2. The paper (Eulerho 1736) is available from the Euler Archive (http://eulerarchive.maa.org), maintained by the Mathematical Association of America. For an English translation, see Biggs, Lloyd, and Wilson 1986.

3. The literature on graphs is vast, as is the subject itself. For a good starting point, see Benjamin, Chartrand, and Zhang 2015.

4. Image from the original publication (Eulerho 1736) retrieved from the Wikipedia Commons at https://commons.wikimedia.org/wiki/File:Solutio_problematis_ad_geometriam_situs_pertinentis,_Fig._1.png. The image is in the public domain.

5. Image from Kekulé 1872, retrieved from the Wikipedia at https://en.wikipedia.org/wiki/Benzene#/media/File:Historic_Benzene_Formulae_Kekul%C3%A9_(original).png. The image is in the public domain.

6. For the original publication in German see Hierholzer 1873.

7. For more details on Hierholzer's algorithm and other algorithms for Eulerian paths, see Fleischner 1991. For the use of graphs in genome assembly, see Pevzner, Tang, and Waterman 2001; Compeau, Pevzner, and Tesler 2011.

8. For an analysis of the optimality of the greedy algorithm for online edge coloring, as well as the example of the starlike graph to show the worst case, see Bar-Noy, Motwani, and Naor 1992.

9. In the original fable, the two characters are an ant and cicada. These two characters also feature in Latin translations of the original ancient Greek and Jean de La Fontaine's retelling of the fable in French.

10. The invention episode is recounted by Dijkstra in his interview in Misa and Frana 2010.

Chapter 3

1. For the first description of the Matthew effect, see Merton 1968. For overviews of the range of phenomena manifesting unequal distributions, see Barabási and Márton 2016; West 2017. For the stadium height and wealth disparity, see Taleb 2007.

2. John McCabe (1965) presented a self-organized search. For analyses of the performance of the move-to-front and transposition methods, see Rivest 1976; Bachrach, El-Yaniv, and Reinstädtler 2002.

3. The secretary problem appeared in Martin Gardner's column in February 1960 in *Scientific American*. A solution was given in the March 1960 issue. For its history, see Ferguson 1989. J. Neil Bearden (2006) provided the solution for the not all-or-nothing variant. Matt Parker (2014, chapter 11) presents the problem, along with several other mathematical ideas and an introduction to computers.

4. Binary search goes back to the dawn of the computer age (Knuth 1998). John Mauchly, one of the designers of the ENIAC, the first general-purpose electronic digital computer, described it in 1946. For the checkered history of binary search, see Bentley 2000; Pattis 1988; Bloch 2006.

Chapter 4

1. Hollerith 1894.

2. Selection and insertion sort have been with us since the dawn of computers; they were included in a survey of sorting published in the 1950s (Friend 1956).

3. According to Knuth (1998, 170), the idea behind radix sort that we have seen here seems to have been around at least since the 1920s.

4. Flipping the coin 226 times follows from $1 / 52! \approx (1 / 2)^{226}$. The example of picking an atom from the earth is from David Hand (2014), according to whom probabilities less than one in 10^{50} are negligible on the cosmic scale.

5. See Hoare 1961a, 1961b, 1961c.

6. For more on randomized algorithms, see Mitzenmacher and Upfal 2017.

7. For an account of von Neumann's life and the environment around the origins of digital computers, see Dyson 2012. For a presentation of von Neumann's merge sort program, see Knuth 1970.

Chapter 5

1. The original PageRank algorithm was published by Brin and Page (1998). We glossed over the mathematics used by the algorithm. For a more in-depth treatment, see Bryan and Leise 2006. For an introduction to search engines

and PageRank, see Langville and Meyer 2006; Berry and Browne 2005. Apart from PageRank, another important algorithm used for ranking is Hypertext Induced Topic Search, or HITS (Kleinberg 1998, 1999), developed before PageRank. Similar ideas had been developed in other fields (sociometry, the quantitative study of social relationships, and econometrics, the quantitative study of economic principles) much earlier, going back to the 1940s (Franceschet 2011).

Chapter 6

1. Although today we can use technology to see neurons in much greater detail, Ramón y Cajal was a pioneer, and his drawings rank among the most elegant illustrations in the history of science. You can find neuron images aplenty on the web, but this image is enough for us, and a simple web search should convince you of the beauty and enduring power of Ramón y Cajal's illustrations. The image is in the public domain, retrieved from https://commons.wikimedia.org/wiki/File:PurkinjeCell.jpg.
2. To be accurate, sigmoid would refer to the Greek letter sigma, which is Σ, yet its appearance is closer to the Latin S.
3. The tangent of an angle is defined as the ratio of the opposite side to the adjacent side in a straight triangle, or equivalently, by the sine of the angle divided by the cosine of the angle in the unit circle. The hyperbolic tangent is defined as the ratio of the hyperbolic sine by the hyperbolic cosine of an angle on a hyperbola.
4. Warren McCulloch and Walter Pitts (1943) proposed the first artificial neuron. Frank Rosenblatt (1957) described the Perceptron. If they are more than half a century old, how come neural networks have become all the rage recently? Marvin Minsky and Seymour Papert (1969) struck a major blow to Perceptrons in their famous book of the same name, which showed that a single Perceptron had fundamental computing limitations. This, coupled with the hardware limitations of the time, ushered in a so-called winter in neural computation, which lasted well until the 1980s, when researchers found how to build and train complex neural networks. Interest in the field then revived, but still a lot more work was required to advance neural networks to the media-grabbing results that we have been seeing in the last few years.
5. One of the challenges in neural networks is that the notation can be off-putting and hence the material seems approachable only to the initiated. In fact, it is not that complicated once you know what it is about. You often

see derivatives; the derivative of a function $f(x)$ with respect to x is written $\frac{df(x)}{dx}$. The partial derivative of a function f of many variables, say, $x_1, x_2, \ldots,$ x_n, is written $\frac{\partial f}{\partial x_i}$. The gradient is written $\nabla f = (\frac{\partial f}{\partial x_1},\ldots,\frac{\partial f}{\partial x_n})$.

6. The backpropagation algorithm came onto the scene in the mid-1980s (Rumelhart, Hinton, and Williams 1986), although various derivations of it had appeared back in the 1960s.

7. This image is from the Fashion-MNIST data (Xiao, Rasul, and Vollgraf 2017), which was developed as a benchmark data set for machine learning. This section was inspired by the basic classification TensorFlow tutorial at https://www.tensorflow.org/tutorials/keras/basic_classification.

8. For a description of the first system to beat the Go human champion, see Silver et al. 2016. For an improved system that does not require human knowledge in the form of previously played games, see Silver et al. 2017.

9. The literature on deep learning is vast. For a comprehensive introduction to the topic, see Goodfellow, Bengio, and Courville 2016. For a shorter and more approachable treatment, see Charniak 2018. For a concise overview, see LeCun, Bengio, and Hinton 2015. For deep and machine learning, see Alpaydin 2016. For a survey of automated neural architecture search methods, see Elsken, Hendrik Metzen, and Hutter 2018.

Epilogue

1. Besides Turing, other names on the short list were Mary Anning, Paul Dirac, Rosalind Franklin, William Herschel and Caroline Herschel, Dorothy Hodgkin, Ada Lovelace and Charles Babbage, Stephen Hawking, James Clerk Maxwell, Srinivasa Ramanujan, Ernest Rutherford, and Frederick Sanger. Babbage, Lovelace, and Turing were all computer pioneers. Babbage (1791–1871) invented the first mechanical computer and developed the essential ideas of modern computers. Lovelace (1815–1852), the daughter of Lord Byron, worked with Babbage, recognized the potential of his invention, and was the first to develop an algorithm that would run on such a machine. She is now considered to have been the first computer programmer. For the £50 design, see the official announcement at https://www.bankofengland.co.uk/news/2019/july/50-pound-banknote-character-announcement.

2. See the excellent biography by Andrew Hodges (1983). Turing's role in breaking the German Enigma cryptographic machine were dramatized in the 2014 film *The Imitation Game*.

3. For a description of the machine, see Turing 1937, 1938.

4. The Turing machine example is adapted from John Hopcroft, Rajeev Motwani, and Jeffrey Ullman (2001, chapter 8). The figure is based on Sebastian Sardina's example at http://www.texample.net/tikz/examples/turing-machine-2/.

5. For more on the Church-Turing thesis, see Lewis and Papadimitriou 1998, chapter 5. For a discussion of the history of the Church-Turing thesis and various variants, see Copeland and Shagrir 2019.

REFERENCES

Alpaydin, Ethem. 2016. *Machine Learning*. Cambridge, MA: MIT Press.

Bachrach, Ran, Ran El-Yaniv, and Martin Reinstädtler. 2002. "On the Competitive Theory and Practice of Online List Accessing Algorithms." *Algorithmica* 32 (2): 201–245.

Barabási, Albert-László, and Pósfai Márton. 2016. *Network Science*. Cambridge: Cambridge University Press.

Bar-Noy, Amotz, Rajeev Motwani, and Joseph Naor. 1992. "The Greedy Algorithm Is Optimal for Online Edge Coloring." *Information Processing Letters* 44 (5): 251–253.

Bearden, J. Neil. 2006. "A New Secretary Problem with Rank-Based Selection and Cardinal Payoffs." *Journal of Mathematical Psychology* 50:58–59.

Benjamin, Arthur, Gary Chartrand, and Ping Zhang. 2015. *The Fascinating World of Graph Theory*. Princeton, NJ: Princeton University Press.

Bentley, Jon. 2000. *Programming Pearls*. 2nd ed. Boston: Addison-Wesley.

Berry, Michael W., and Murray Browne. 2005. *Understanding Text Engines: Mathematical Modeling and Text Retrieval*. 2nd ed. Philadelphia: Society for Industrial and Applied Mathematics.

Biggs, Norman L., E. Keith Lloyd, and Robin J. Wilson. 1986. *Graph Theory, 1736–1936*. Oxford: Clarendon Press.

Bjorklund, Eric. 1999. "The Theory of Rep-Rate Pattern Generation in the SNS Timing System." SNS-NOTE-CNTRL-99. Spallation Neutron Source. https://ics-web.sns.ornl.gov/timing/Rep-Rate%20Tech%20Note.pdf.

Bloch, Joshua. 2006. "Extra, Extra—Read All about It: Nearly All Binary Searches and Mergesorts Are Broken." *Google AI Blog*, June 2. http://googleresearch.blogspot.it/2006/06/extra-extra-read-all-about-it-nearly.html.

Brin, Sergey, and Lawrence Page. 1998. "The Anatomy of a Large-Scale Hypertextual Web Search Engine." *Computer Networks and ISDN Systems* 30 (1–7): 107–117.

Bryan, Kurt, and Tanya Leise. 2006. "The $25,000,000,000 Eigenvector: The Linear Algebra behind Google." *SIAM Review* 48 (3): 569–581.

Charniak, Eugene. 2018. *Introduction to Deep Learning*. Cambridge, MA: MIT Press.

Compeau, Phillip E. C., Pavel A. Pevzner, and Glenn Tesler. 2011. "How to Apply de Bruijn Graphs to Genome Assembly." *Nature Biotechnology* 29 (11): 987–991.

Copeland, B. Jack, and Oron Shagrir. 2019. "The Church-Turing Thesis: Logical Limit or Breachable Barrier?" *Communications of the ACM* 62 (1): 66–74.

Demaine, Erik D., Francisco Gomez-Martin, Henk Meijer, David Rappaport, Perouz Taslakian, Godfried T. Toussaint, Terry Winograd, and David R. Wood. 2009. "The Distance Geometry of Music." *Computational Geometry: Theory and Applications* 42 (5): 429–454.

Dyson, George. 2012. *Turing's Cathedral: The Origins of the Digital Universe*. New York: Vintage Books.

Elsken, Thomas, Jan Hendrik Metzen, and Frank Hutter. 2018. "Neural Architecture Search: A Survey." ArXiv, Cornell University. August 16. http://arxiv.org/abs/1808.05377.

Eulerho, Leonhardo. 1736. "Solutio Problematis Ad Geometrian Situs Pertinentis." *Commetarii Academiae Scientiarum Imperialis Petropolitanae* 8:128–140.

Ferguson, Thomas S. 1989. "Who Solved the Secretary Problem?" *Statistical Science* 4 (3): 282–289.

Fleischner, Herbert, ed. 1991. "Chapter X Algorithms for Eulerian Trails and Cycle Decompositions, Maze Search Algorithms." In *Eulerian Graphs and Related Topics*, 50:X.1–X.34. Amsterdam: Elsevier.

Franceschet, Massimo. 2011. "PageRank: Standing on the Shoulders of Giants." *Communications of the ACM* 54 (6): 92–101.

Friend, Edward H. 1956. "Sorting on Electronic Computer Systems." *Journal of the ACM* 3 (3): 134–168.

Goodfellow, Ian, Yoshua Bengio, and Aaron Courville. 2016. *Deep Learning*. Cambridge, MA: MIT Press.

Hand, David J. 2014. *The Improbability Principle: Why Coincidences, Miracles, and Rare Events Happen Every Day*. New York: Farrar, Straus and Giroux.

Hawking, Stephen. 1988. *A Brief History of Time*. New York: Bantam Books.

Hierholzer, Carl. 1873. "Ueber die Möglichkeit, einen Linienzug ohne Wiederholung und ohne Unterbrechung zu Umfahren." *Mathematische Annalen* 6 (1): 30–32.

Hoare, C. A. R. 1961a. "Algorithm 63: Partition." *Communications of the ACM* 4 (7): 321.

Hoare, C. A. R. 1961b. "Algorithm 64: Quicksort." *Communications of the ACM* 4 (7): 321.

Hoare, C. A. R. 1961c. "Algorithm 65: Find." *Communications of the ACM* 4 (7): 321–322.

Hodges, Andrew. 1983. *Alan Turing: The Enigma*. New York: Simon and Schuster.

Hollerith, Herman. 1894. "The Electrical Tabulating Machine." *Journal of the Royal Statistical Society* 57 (4): 678–689.

Hopcroft, John E., Rajeev Motwani, and Jeffrey D. Ullman. 2001. *Introduction to Automata Theory, Languages, and Computation*. 2nd ed. Boston: Addison-Wesley.

Iwashita, Hiroaki, Yoshio Nakazawa, Jun Kawahara, Takeaki Uno, and Shin-ichi Minato. 2013. "Efficient Computation of the Number of Paths in a Grid Graph with Minimal Perfect Hash Functions." Technical Report TCS-TR-A-13-64. Division of Computer Science, Graduate School of Information Science, Technology, Hokkaido University.

Kekulé, August. 1872. "Ueber Einige Condensationsprodukte Des Aldehyds." *Annalen der Chemie und Pharmacie* 162 (1): 77–124.

Kleinberg, Jon M. 1998. "Authoritative Sources in a Hyperlinked Environment." In *Proceedings of the Ninth Annual ACM-SIAM Symposium on Discrete Algorithms*, 668–677. Philadelphia: Society for Industrial and Applied Mathematics.

Kleinberg, Jon M. 1999. "Authoritative Sources in a Hyperlinked Environment." *Journal of the ACM* 46 (5): 604–632.

Knuth, Donald E. 1970. "Von Neumann's First Computer Program." *Computing Surveys* 2 (4): 247–261.

Knuth, Donald E. 1972. "Ancient Babylonian Algorithms." *Communications of the ACM* 15 (7): 671–677.

Knuth, Donald E. 1997. *The Art of Computer Programming, Volume 1: Fundamental Algorithms*. 3rd ed. Reading, MA: Addison-Wesley.

Knuth, Donald E. 1998. *The Art of Computer Programming, Volume 3: Sorting and Searching*. 2nd ed. Reading, MA: Addison-Wesley.

Knuth, Donald E. 2011. *The Art of Computer Programming, Volume 4A: Combinatorial Algorithms, Part 1*. Upper Saddle River, NJ: Addison-Wesley.

Langville, Amy N., and Carl D. Meyer. 2006. *Google's PageRank and Beyond: The Science of Search Engine Rankings*. Princeton, NJ: Princeton University Press.

LeCun, Yann, Yoshua Bengio, and Geoffrey Hinton. 2015. "Deep Learning." *Nature* 521 (7553): 436–444.

Lewis, Harry R., and Christos H. Papadimitriou. 1998. *Elements of the Theory of Computation*. 2nd ed. Upper Saddle River, NJ: Prentice Hall.

McCabe, John. 1965. "On Serial Files with Relocatable Records." *Operations Research* 13 (4): 609–618.

McCulloch, Warren S., and Walter Pitts. 1943. "A Logical Calculus of the Ideas Immanent in Nervous Activity." *Bulletin of Mathematical Biophysics* 5 (4): 115–133.

Merton, Robert K. 1968. "The Matthew Effect in Science." *Science* 159 (3810): 56–63.

Minsky, Marvin, and Seymour Papert. 1969. *Perceptrons: An Introduction to Computational Geometry*. Cambridge, MA: MIT Press.

Misa, Thomas J., and Philip L. Frana. 2010. "An Interview with Edsger W. Dijkstra." *Communications of the ACM* 53 (8): 41–47.

Mitzenmacher, Michael, and Eli Upfal. 2017. *Probability and Computing: Randomization and Probabilistic Techniques in Algorithms and Data Analysis*. 2nd ed. Cambridge: Cambridge University Press.

Parker, Matt. 2014. *Things to Make and Do in the Fourth Dimension: A Mathematician's Journey through Narcissistic Numbers, Optimal Dating Algorithms, at Least Two Kinds of Infinity, and More*. London: Penguin Books.

Pattis, Richard E. 1988. "Textbook Errors in Binary Searching." *SIGCSE Bulletin* 20 (1): 190–194.

Pevzner, Pavel A., Haixu Tang, and Michael S. Waterman. 2001. "An Eulerian Path Approach to DNA Fragment Assembly." *Proceedings of the National Academy of Sciences* 98 (17): 9748–9753.

Pinker, Steven. 2018. *Enlightenment Now: The Case for Reason, Science, Humanism, and Progress*. New York: Viking Press.

Rivest, Ronald. 1976. "On Self-Organizing Sequential Search Heuristics." *Communications of the ACM* 19 (2): 63–67.

Rosenblatt, Frank. 1957. "The Perceptron: A Perceiving and Recognizing Automaton." Report 85-460-1. Cornell Aeronautical Laboratory.

Rumelhart, David E., Geoffrey E. Hinton, and Ronald J. Williams. 1986. "Learning Representations by Back-Propagating Errors." *Nature* 323 (6088): 533–536.

Silver, David, Aja Huang, Chris J. Maddison, Arthur Guez, Laurent Sifre, George van den Driessche, Julian Schrittwieser, et al. 2016. "Mastering the Game of Go with Deep Neural Networks and Tree Search." *Nature* 529 (7587): 484–489.

Silver, David, Julian Schrittwieser, Karen Simonyan, Ioannis Antonoglou, Aja Huang, Arthur Guez, Thomas Hubert, et al. 2017. "Mastering the Game of Go without Human Knowledge." *Nature* 550 (7676): 354–359.

Strogatz, Steven. 2012. *The Joy of x: A Guided Tour of Math, from One to Infinity*. New York: Houghton Mifflin Harcourt.

Taleb, Nassim Nicholas. 2007. *The Black Swan: The Impact of the Highly Improbable*. New York: Random House.

Toussaint, Godfried T. 2005. "The Euclidean Algorithm Generates Traditional Musical Rhythms." In *Renaissance Banff: Mathematics, Music, Art, Culture*, edited by Reza Sarhangi and Robert V. Moody, 47–56. Winfield, KS: Bridges Conference, Southwestern College.

Toussaint, Godfried T. 2013. *The Geometry of Musical Rhythm: What Makes a "Good" Rhythm Good?* Boca Raton, FL: CRC Press.

Turing, Alan M. 1937. "On Computable Numbers, with an Application to the Entscheidungsproblem." *Proceedings of the London Mathematical Society* S2–42:230–265.

Turing, Alan M. 1938. "On Computable Numbers, with an Application to the Entscheidungsproblem. A Correction." *Proceedings of the London Mathematical Society* S2–43:544–546.

Tyson, Neil deGrasse, Michael Abram Strauss, and Richard J. Gott. 2016. *Welcome to the Universe: An Astrophysical Tour*. Princeton, NJ: Princeton University Press.

West, Geoffrey. 2017. *Scale: The Universal Laws of Life, Growth, and Death in Organisms, Cities, and Companies*. London: Weidenfeld and Nicholson.

Xiao, Han, Kashif Rasul, and Roland Vollgraf. 2017. "Fashion-MNIST: A Novel Image Dataset for Benchmarking Machine Learning Algorithms." August 28. https://arxiv.org/abs/1708.07747.

FURTHER READING

Broussard, Meredith. 2018. *Artificial Unintelligence: How Computers Misunderstand the World*. Cambridge, MA: MIT Press.

Christian, Brian, and Tom Griffiths. 2016. *Algorithms to Live By: The Computer Science of Human Decisions*. New York: Henry Holt and Company.

Cormen, Thomas H. 2013. *Algorithms Unlocked*. Cambridge, MA: MIT Press.

Cormen, Thomas H., Charles E. Leiserson, Ronald L. Rivest, and Clifford Stein. 2009. *Introduction to Algorithms*. 3rd ed. Cambridge, MA: MIT Press.

Denning, Peter J., and Matti Tedre. 2019. *Computational Thinking*. Cambridge, MA: MIT Press.

Dewdney, A. K. 1993. *The (New) Turing Omnibus: 66 Excursions in Computer Science*. New York: W. H. Freeman and Company.

Dyson, George. 2012. *Turing's Cathedral: The Origins of the Digital Universe*. New York: Vintage Books.

Erwig, Martin. 2017. *Once upon an Algorithm: How Stories Explain Computing*. Cambridge, MA: MIT Press.

Fry, Hannah. 2018. *Hello World: How to Be Human in the Age of the Machine*. London: Doubleday.

Harel, David, and Yishai Feldman. 2004. *Algorithmics: The Spirit of Computing*. 3rd ed. Harlow, UK: Addison-Wesley.

Louridas, Panos. 2017. *Real-World Algorithms: A Beginner's Guide*. Cambridge, MA: MIT Press.

MacCormick, John. 2013. *Nine Algorithms That Changed the Future: The Ingenious Ideas That Drive Today's Computers*. Princeton, NJ: Princeton University Press.

O'Neil, Cathy. 2016. *Weapons of Math Destruction: How Big Data Increases Inequality and Threatens Democracy*. New York: Crown Publishing Group.

Petzold, Charles. 2008. *The Annotated Turing: A Guided Tour through Alan Turing's Historic Paper on Computability and the Turing Machine*. Indianapolis: Wiley Publishing.

Sedgewick, Robert, and Kevin Wayne. 2017. *Computer Science: An Interdisciplinary Approach*. Boston: Addison-Wesley.

INDEX

The MIT Press Essential Knowledge Series

AI Ethics, Mark Coeckelbergh
Algorithms, Panos Louridas
Anticorruption, Robert I. Rotberg
Auctions, Timothy P. Hubbard and Harry J. Paarsch
The Book, Amaranth Borsuk
Carbon Capture, Howard J. Herzog
Citizenship, Dimitry Kochenov
Cloud Computing, Nayan B. Ruparelia
Collaborative Society, Dariusz Jemielniak and Aleksandra Przegalinska
Computational Thinking, Peter J. Denning and Matti Tedre
Computing: A Concise History, Paul E. Ceruzzi
The Conscious Mind, Zoltan E. Torey
Contraception, Donna Drucker
Critical Thinking, Jonathan Haber
Crowdsourcing, Daren C. Brabham
Cynicism, Ansgar Allen
Data Science, John D. Kelleher and Brendan Tierney
Deep Learning, John D. Kelleher
Extraterrestrials, Wade Roush
Extremism, J. M. Berger
Fake Photos, Hany Farid
fMRI, Peter A. Bandettini
Food, Fabio Parasecoli
Free Will, Mark Balaguer
The Future, Nick Montfort
GPS, Paul E. Ceruzzi
Haptics, Lynette A. Jones
Information and Society, Michael Buckland
Information and the Modern Corporation, James W. Cortada
Intellectual Property Strategy, John Palfrey
The Internet of Things, Samuel Greengard
Irony and Sarcasm, Roger Kreuz
Machine Learning: The New AI, Ethem Alpaydin
Machine Translation, Thierry Poibeau
Macroeconomics, Felipe Larraín B.
Memes in Digital Culture, Limor Shifman
Metadata, Jeffrey Pomerantz

The Mind–Body Problem, Jonathan Westphal
MOOCs, Jonathan Haber
Neuroplasticity, Moheb Costandi
Nihilism, Nolen Gertz
Open Access, Peter Suber
Paradox, Margaret Cuonzo
Post-Truth, Lee McIntyre
Quantum Entanglement, Jed Brody
Recommendation Engines, Michael Schrage
Recycling, Finn Arne Jørgensen
Robots, John Jordan
School Choice, David R. Garcia
Self-Tracking, Gina Neff and Dawn Nafus
Sexual Consent, Milena Popova
Smart Cities, Germaine R. Halegoua
Spaceflight, Michael J. Neufeld
Spatial Computing, Shashi Shekhar and Pamela Vold
Sustainability, Kent E. Portney
Synesthesia, Richard E. Cytowic
The Technological Singularity, Murray Shanahan
3D Printing, John Jordan
Understanding Beliefs, Nils J. Nilsson
Virtual Reality, Samuel Greengard
Waves, Frederic Raichlen

PANOS LOURIDAS is Associate Professor in the Department of Management Science and Technology at the Athens University of Economics and Business. He works on algorithmic applications, software engineering, security, practical cryptography, and applied machine learning. He is the author of *Real-World Algorithms: A Beginners Guide*, published by the MIT Press. He has been an active programmer for more than a quarter of a century.